Lecture Notes in Mathematics

A collection of informal reports and seminars
Edited by A. Dold, Heidelberg and B. Eckmann, Zürich

T0192080

203

Donald Knutson

Columbia University in the City of New York,
New York, NY/USA

Algebraic Spaces

Springer-Verlag
Berlin · Heidelberg · New York 1971

AMS Subject Classifications (1970): 14-02, 14A15, 14A20, 14F20, 18F10

ISBN 3-540-05496-0 Springer-Verlag Berlin · Heidelberg · New York
ISBN 0-387-05496-0 Springer-Verlag New York · Heidelberg · Berlin

Offsetdruck: Julius Beltz, Hemsbach

PREFACE

The core of this book is the author's thesis, Algebraic Spaces, written under Michael Artin at the Massachusetts Institute of Technology. The object there as here was to work out the foundations à la EGA for the theory of algebraic spaces, and hence give the necessary background for Artin's fundamental papers Algebraization of Formal Moduli I, II.

While working on this book, I was supported by M.I.T., Boston College, Columbia University, and the Advanced Science Summer Seminar at Bowdoin College, sponsored by the National Science Foundation. To all these institutions, I extend my gratitude. My special thanks goes to Professor Michael Artin both for many helpful discussions and for his initial suggestion that I undertake this project.

Donald Knutson

CONTENTS

Introduction 1

Chapter One: The Etale Topology of Schemes 29

1. Grothendieck Topologies and Descent Theory 29
2. The Zariski Topology of Schemes 38
3. The Flat Topology of Schemes 52
4. The Etale Topology of Schemes 59
5. Etale Equivalence Relations 72

Chapter Two: Algebraic Spaces 91

1. The Category of Algebraic Spaces 91
2. The Etale Topology of Algebraic Spaces 101
3. Descent Theory for Algebraic Spaces 106
4. Quasicoherent Sheaves and Cohomology 113
5. Local Constructions 120
6. Points and the Zariski Topology 129
7. Proper and Projective Morphisms 139
8. Integral Algebraic Spaces 144

Chapter Three: Quasicoherent Sheaves on Noetherian Locally Separated Algebraic Spaces 153

1. The Completeness / Extension Lemma 153
2. The Serre Criterion 159
3. Schemehood and Nilpotents 165
4. Chevalley's Theorem 169
5. Devissage 173

Chapter Four: The Finiteness Theorem 176

1. Actions of a Finite Group 177
2. Symmetric powers of Projective Spaces 185
3. Chow's Lemma 192
4. The Finiteness Theorem 202

Chapter Five: Formal Algebraic Spaces 204

1. Affine Formal Schemes 204
2. Formal Algebraic Spaces 215
3. The Theorem of Holomorphic Functions 224
4. Applications to Proper Morphisms 233
5. Completions of Modules of Homomorphisms 241
6. The Grothendieck Existence Theorem 245

Index 252

Bibliography 257

INTRODUCTION

The notion of algebraic space is a generalization of the notion of scheme, as presented in A. Grothendieck's *Elements de Geometrie Algebrique* (EGA). In this introduction, we give a tentative definition of algebraic space, a number of examples, and some indication of the literature on the subject.

The rest of the book is an orderly development of the foundations of the theory. Chapter I contains the relevant background with all the definitions (but not proofs) of scheme theory and descent theory. This background is drawn from EGA, M. Artin's *Grothendieck Topologies* (GT), and the various *Seminaires de Geometrie Algebrique* (SGA, SGAA, and SGAD) of Grothendieck, Artin, P. Deligne, J.L. Verdier, and M. Demazure. The reader who is vaguely familiar with this material, i.e., the reader who feels happy with the phrase "etale topology of schemes", can skip Chapter I except for a brief reading of sections 1 and 5.

Algebraic spaces arise in the attempt to construct certain algebraic-geometric objects: Hilbert Schemes, Picard Schemes, and moduli varieties; and in the attempt to impose algebraic structure on certain given objects, such as analytic spaces.

As an example, we consider the problem of constructing the Hilbert Scheme H of a complex algebraic variety X. Ideally, H should be a scheme, the closed points of which parametrize the closed subvarieties of X in a continuous manner. This requirement does not determine H uniquely, so we impose a naturality condition: for any scheme S, $\mathrm{Hom}_{\mathrm{Schemes}}(S,H)$ should be the set of families of closed subvarieties of X, parametrized by S. Specifically, we define for each locally noetherian scheme S, $\mathrm{Hilb}_X(S) = \{$Set of closed subschemes Z of $X \times S$ such that the map $Z \to S$ is flat and proper.$\}$

$\mathrm{Hilb}_X(S)$ is a contravariant functor from the category of locally noetherian schemes to the category of sets. We now define: a variety X <u>has a Hilbert Scheme</u> if and only if the functor $\mathrm{Hilb}_X(-)$ is representable. If $\mathrm{Hilb}_X(-)$ is representable by a scheme H, i.e., if the functors $\mathrm{Hilb}_X(-)$ and $\mathrm{Hom}_{(\mathrm{Schemes})}(-,H)$ are equivalent, then H is uniquely determined and is the <u>Hilbert Scheme of X</u>.

In this way we view the problem of constructing a certain object as the problem of representing a certain functor.

In the specific case of Hilbert Schemes, it has been shown (XVII) that if X is a projective variety, then X has a Hilbert Scheme. In general, however, an arbitrary variety need not have a Hilbert Scheme (see Example 3 below).

To find objects which might serve as Hilbert Schemes, we must look at a larger category than that of schemes. Because of the results of Artin (II,III,IV,V), described briefly below, it seems likely that the notion of algebraic space provides the appropriate larger category. (It should be mentioned here that Douady has shown the representability of the Hilbert Functor in the category of analytic spaces [XII].)

Our first requirement for algebraic spaces is:

I. The category of algebraic spaces contains the category of schemes as a faithful and full subcategory

A second requirement comes from the topology of schemes. The category of schemes has several Grothendieck topologies, among others the Zariski topology, the etale topology, and the flat topology. In any of these, given a scheme X, the functor $X^{\cdot} = \mathrm{Hom}(-,X):(\mathrm{Schemes}) \to (\mathrm{Sets})$ satisfies the sheaf axiom.[1] Hence, to represent a functor like $\mathrm{Hilb}_X(-)$, one must know that the functor at least satisfies the sheaf axiom. On the other hand, from the uniqueness involved in representing functors, every scheme is uniquely determined by its associated

[1] Throughout we denote by (- - -) the category of all - - -. Also, we use the notation $X^{\cdot}$ for the contravariant functor represented by X.

sheaf. A sheaf is determined by its values on affine schemes and by the Yoneda Lemma, every map of representable sheaves $X^{\cdot} \to Y^{\cdot}$ is induced by a unique map $X \to Y$ of schemes. Thus there are full faithful imbeddings

$$(\text{Affine Schemes}) \hookrightarrow (\text{Schemes}) \hookrightarrow (\text{Sheaves}) \quad .$$

To carry this over to algebraic spaces, we must first pick the appropriate Grothendieck topology on the category of schemes. For several reasons, we choose the etale topology. On one hand, for schemes over the complex numbers, this topology is defined much like the usual analytic topology. In contrast to finer topologies such as the flat topology, the notions of point and reduced scheme are stable in the etale topology. On the other hand, there are indications (e.g., Grothendieck's work on the Weil Conjectures) that the Zariski topology is not fine enough. Finally, the etale topology is the context in which Artin has proved his above-mentioned theorems. Hence we take as our second requirement for algebraic spaces:

II. <u>The category of algebraic spaces has a Grothendieck topology, whose restriction to the subcategory of schemes is the etale topology.</u> <u>For every algebraic space</u> X, <u>the functor</u> $X^{\cdot} = \mathrm{Hom}_{(\mathrm{Alg.Spaces})}(-,X)$: (Algebraic Spaces) $\to$ (Sets) *is a sheaf and the transformation* $X \rightsquigarrow X^{\cdot}$ *gives a full and faithful imbedding* (Algebraic Spaces) → (Sheaves of sets on schemes in the etale topology).

We might stop here and just define an algebraic space to be a sheaf. But to do much geometry, we need to mimic another aspect of scheme theory which is, roughly, that any property or construction on the category of all schemes (such as the giving of a sheaf) is determined by the property or construction on the subcategory of affine schemes. To make this aspect more clear, we go back to the case of schemes in the Zariski topology.

If X is a scheme (quasicompact and separated, say) and $\{U_i \to X\}_{i=1,2,\ldots,n}$ is an open covering of X by affine schemes, then X is completely described by giving just the family $\{U_i\}$ and the "gluing data", the open subsets $V_{ij} = U_i \cap U_j$ of X and the immersions $V_{ij} \to U_i$ and $V_{ij} \to U_j$. More abstractly, we can write U for the disjoint union of the U_i's. Noting that $V_{ij} = U_i \times_X U_j$, we can write $R = U \times_X U$ for the disjoint union of the V_{ij}'s.

There is a canonical injection $R \to U \times U$ which identifies R as an equivalence relation on U. Let π_1 and π_2 be the two projections $R \rightrightarrows U$ and $\pi : U \to X$ the covering map. Then in the diagram

$$R \underset{\pi_2}{\overset{\pi_1}{\rightrightarrows}} U \xrightarrow{\ \pi\ } X$$

$\pi: U \to X$ is the cokernel of the maps π_1, π_2 in the category of schemes. There is also a canonical injection $U \to R$ whose image is a component of R, isomorphic to U, the <u>diagonal component</u>.

R and U are both affine schemes (if X is quasicompact and separated) and thus X is described by an equivalence relation on affine schemes. In this way schemes are constructed by "adding quotients of reasonable equivalence relations" to the category of affine schemes.

But one must be careful here. The equivalence relation $R \rightrightarrows U$ of affine schemes may already have a quotient in the category of affine schemes, but it may be the "wrong" quotient. E.g., if X is a complete variety over a closed field k, the affine quotient of $R \rightrightarrows U$ will be Spec k, not X.

In scheme theory, this difficulty is circumvented by taking X as the quotient of $R \rightrightarrows U$ in the category of local ringed spaces. Tying this in with the considerations above, one then shows that the functor $X^{\cdot}$ is a sheaf on affine schemes in the Zariski topology and that $X^{\cdot}$ is the quotient, in the category of sheaves, of the equivalence relation $R^{\cdot} \rightrightarrows U^{\cdot}$.

The point of this equivalence relation construction is that all the information and constructions on X--quasicoherent sheaves, closed subvarieties, etc.--is determined locally on U modulo gluing data on R.

Etale descent theory (Chapter I) shows that all the information is equally determined by giving an arbitrary etale surjective map $\pi : U \to X$ with $R = U \times_X U$. Here X' is the quotient of $R' \rightrightarrows U'$ in the category of sheaves in the etale topology.

In either of these cases, there is a cartesian diagram

$$\begin{array}{ccc} R & \xrightarrow{\delta} & U \times U \\ \downarrow & & \downarrow {\scriptstyle \pi \times \pi} \\ X & \xrightarrow{\Delta} & X \times X \end{array}$$

where $\pi \times \pi$ is an etale surjective map. Applying descent theory, we see that δ has the same properties as Δ and so separation conditions on X are imposed by making restrictions on δ.

With all this in mind, we take as a final requirement for algebraic spaces:

III. *For each algebraic space* X, *there is a scheme* U *and a covering map* $U \to X$ *such that the fiber product* $R = U \times_X U$ *is a scheme, the maps* $R \rightrightarrows U$ *are etale surjective, and the map* $R \to U \times U$ *is quasicompact*. X *is locally separ-*

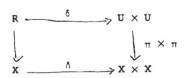

ated if $R \to U \times U$ is a quasicompact immersion, and separated if $R \to U \times U$ is a closed immersion.

These three requirements more or less determine the formal definition of algebraic space, which is given in II.1.1.[1]

One might be tempted here to construct X as the local ringed space quotient X_1 of $R \rightrightarrows U$. Unfortunately, this can sometimes be the wrong quotient, in the sense that R might not be $U \times_{X_1} U$. (See Example 1 below.)

The quasicompactness in III is a technical requirement to make the general foundations neater. Equivalent requirements are given in I.5.12.

There are schemes X for which the diagonal map $X \to X \times X$ is not quasicompact (so-called non-quasi-separated schemes). We choose to regard these as pathological examples and assume throughout the book that all schemes considered are quasi-separated.

In Chapter II, the formal development of the theory of algebraic spaces begins and the rest of the book is devoted to showing that algebraic spaces are very much like schemes. In the rest of this Introduction, we give examples to show in what ways algebraic spaces are not like schemes. We also

---

[1] Throughout we use the notation A.n.m to denote paragraph m of section n of Chapter A. Within A, A.n.m is written just n.m.

make some general comments on the relation of scheme theory to algebraic space theory.

Example 0. Every (quasiseparated) scheme is an algebraic space.

Example 1. An algebraic space which is not locally separated and not a scheme.

In this example, we work with schemes defined over the complex numbers. Let U be the scheme obtained by taking two copies of the affine line A_1 = Spec $\mathbf{C}[s]$, A_2 = Spec $\mathbf{C}[t]$ and identifying the points $s = 0$ and $t = 0$.[1] We denote the common point by p. The real points of U then look like a figure  with p the center point. Let R consist of one copy of U, the diagonal part of R, and a scheme U' obtained from U by deleting the point p.

In pictures, $R \rightrightarrows U$ is

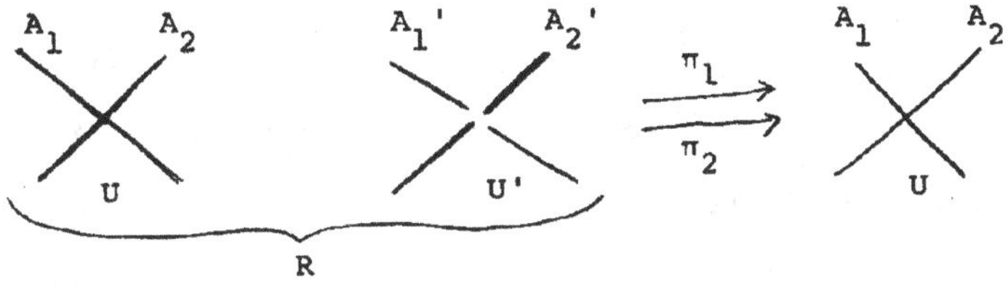

π_1 and π_2 are the identity maps on $U \subset R$ and on U', $\pi_1 : A'_i \to A_i$, $i = 1,2$ and $\pi_2 : A'_i \to A_{3-i}$, $i = 1,2$. This is an

---

[1] Throughout the book, $\mathbf{C}$ denotes the complex numbers, $\mathbf{Q}$ the rationals, and $\mathbf{Z}$ the ring of integers.

etale equivalence relation which is quasiseparated but not locally separated.

The quotient of $R \rightrightarrows U$ in the category of local ringed spaces is clearly the affine line A. But $R \neq U \times_A U$. Thus the algebraic space X which is the quotient of $R \rightrightarrows U$ in the category of sheaves is not a local ringed space (and hence in particular not a scheme). In fact, X looks like a copy of the affine line A = Spec C[s] except at the point s = 0 where where X has, not a double point, but a single point with two tangent directions:

X: 

s = 0

<u>Example 2</u>. <u>Locally Separated Algebraic Spaces.</u>

We start with a general construction. Let X and U be schemes and $\pi: U \to X$ an etale surjective quasifinite map. Let $R = U \times_X U$. Then, applying descent theory, X is the quotient of the etale equivalence relation $R \rightrightarrows U$.

Now let T be a subscheme of X and $i: T \to X$ the immersion. We write $U_T = T \times_X U$ and $R_T = R \times_X U$ and a simple diagram chase shows that $R_T = U_T \times_T U_T$ and that $R_T \rightrightarrows U_T$ is an etale equivalence relation with quotient T:

$$\begin{array}{ccccc} R_T & \rightrightarrows & U_T & \longrightarrow & T \\ \downarrow & & \downarrow & & \downarrow \\ R & \rightrightarrows & U & \longrightarrow & X \end{array}$$

This construction works in general, with X an arbitrary algebraic space defined by an etale equivalence relation $R \rightrightarrows U$. Below we use this to calculate some of the properties of X.

Going back to the case where X is a scheme, the quotient of $R \rightrightarrows U$, we now modify R to obtain a new equivalence relation R' on U.

With the notation above, R_T consists of two parts, the diagonal $\Delta \cong U_T$, and another component R'_T. R'_T is a subspace of R_T, so a subspace of R. Let $R' \subset R$ be the complement of R'_T. Then $R' \rightrightarrows U$ is a new etale equivalence relation. Let X' be its quotient and $f: X' \to X$ the canonically induced map:

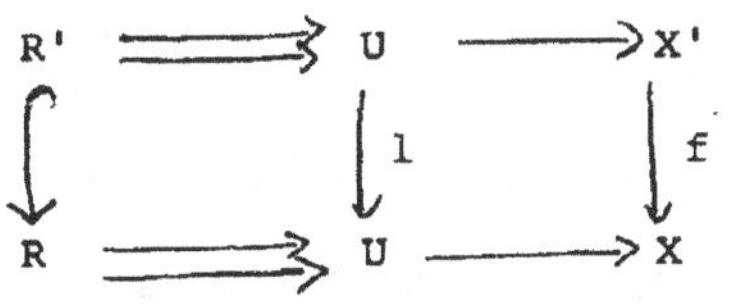

In general f is not an isomorphism. Indeed, the new equivalence relation $R' \rightrightarrows U$ has been constructed to be trivial on the subspace $U_T \to U$ and a bit of diagram chasing shows there is a cartesian diagram

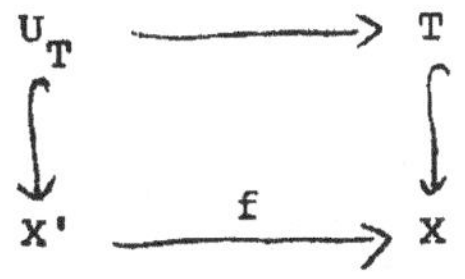

where the vertical arrows are subspaces and the restriction of f, $f_{(X'-U_T)}:(X'-U_T) \to (X-T)$, is an isomorphism.

Thus X' looks like X but with an etale extension along the subspace T.

We now give two specific examples. For the first, let X be obtained from the affine plane over the complex numbers, Spec $\mathbf{C}[P,Q]$ by removing the line $Q = 0$. Let U be the same: Spec $\mathbf{C}[P',Q']$ minus the line $Q' = 0$. Let $U \to X$ be the etale map obtained from the map of rings $\mathbf{C}[P,Q] \to \mathbf{C}[P',Q']$, $P \rightsquigarrow P'$, $Q \rightsquigarrow Q'^2$.

We first let $T_1 \subset X$ be the closed subscheme which is the line $Q = 1$. The inverse image U_{T_1} is Spec $\mathbf{C}[P',Q']/(Q'^2 = 1)$ (minus the line $Q' = 0$) and consists of two disjoint copies of the affine line. The associated X'_1 is the plane, minus the $Q = 0$ axis, with the line $Q = 1$ doubled. Thus X'_1 is a scheme, although not a separated scheme.

Now let $T_2 \to X$ be the closed subscheme which is the line $P = 1$. Then $U_{T_2} = $ Spec $\mathbf{C}[P',Q']/(P' = 1)$ (minus the line $Q' = 0$) and is connected! The associated X'_2 is the plane, minus the $Q = 0$ axis, with the line $P = 1$ replaced by a nontrivial double covering $U_{T_2} \to T_2$. X'_2 is not a scheme since no open subset of X'_2 containing the generic point of U_{T_2} is an affine scheme. But X'_2 is a locally separated algebraic space.

For a second example, we start with Spec $\mathbb{Z}$. It will be convenient to write x_q for the affine spectrum of the finite field containing $q = p^n$ elements. We can draw Spec $\mathbb{Z}$ as follows:

Spec Z: $\quad \overset{\cdot}{x_2} \;\; \overset{\cdot}{x_3} \;\; \overset{\cdot}{x_5} \;\; \overset{\cdot}{x_7} \quad$ Spec Q

Let $X = (\mathrm{Spec}\ Z - x_2)$. Consider the ring extension $\mathbb{Z} \to \mathbb{Z}[\sqrt{2}]$. Let $U = X \underset{\mathrm{Spec}\ \mathbb{Z}}{\times} (\mathrm{Spec}\ \mathbb{Z}[\sqrt{2}])$. Then $U \to X$ is an etale covering.

Given a point $T = x_p \in X$, there are two possibilities for $U_T = T \underset{X}{\times} U$. If the equation $t^2 = 2$ is solvable in $\mathbb{Z}/p\mathbb{Z}$, then U_T is a disjoint union of two copies of x_p. If we perform the general construction above in this case, the new object has the form:

$\overset{\cdot}{x_3} \;\; \overset{\cdot}{x_5} \;\; \overset{\cdot}{x_7} \;\; \cdots \;\; \overset{:}{x_p} \;\; \cdots \quad$ Spec Q

which is a scheme: X with a single point x_p doubled.

On the other hand, if $t^2 = 2$ is not solvable in $\mathbb{Z}/p\mathbb{Z}$, (e.g., $p = 5$), U_T is the single point $x_{(p^2)}$ and performing our construction we get

X': $\quad \overset{\cdot}{x_3} \;\; \overset{\cdot}{x_5} \;\; \overset{\cdot}{x_7} \;\; \cdots \;\; \overset{\cdot}{x_{(p^2)}} \;\; \cdots \quad$ Spec Q which is X with a field extension at one point! Clearly not a scheme.

This X' illustrates another phenomenon of algebraic spaces. Given an algebraic space Y, defined by an etale equivalence relation $S \rightrightarrows V$, we can construct an associated local ringed space $|Y|$ by taking $|Y|$ to be the quotient of $S \rightrightarrows V$ in the category of local ringed spaces. For the algebraic space X' constructed above, $|X'|$ is isomorphic to X. Thus the functor $|\ |$ applied to the map $X' \xrightarrow{f} X$ yields an isomorphism of local ringed spaces. Hence this functor is not full. Similarly, X' has a nontrivial automorphism (the automorphism of $x_{(p^2)}$) which becomes the identity map of $|X'| = X$. Hence $|\ |$ is also not faithful.

One could ask whether the functor $|\ |$ is fully faithful when restricted to separated algebraic spaces. This we do not know. But even if it were, enough nonseparated algebraic spaces occur in nature (see below) and it would not be sufficient to restrict our attention to the separated case.

Example 3. Quotients by Group Actions. A separated nonsingular algebraic space which is not a scheme. Two theorems of Artin.

This example is due to Hironaka (XVIII) and is taken from Mumford (XXVI).

All schemes considered here are defined over the complex numbers.

Let V_0 be projective 3-space and γ_1 and γ_2 two conics intersecting normally in exactly two points P_1 and P_2. For $i = 1,2$, we construct $\overline{V}_i$ by first blowing up γ_i, and then γ_{3-i} in the result. Let V_i be the open set in $\overline{V}_i$ of points laying over $(V_0 - P_{3-i})$. Let U be obtained by patching V_1 and V_2 together along the common open subset.

U is a nonsingular variety on which, over P_1 and P_2, the two curves γ_1 and γ_2 have been blown up in opposite order

U:

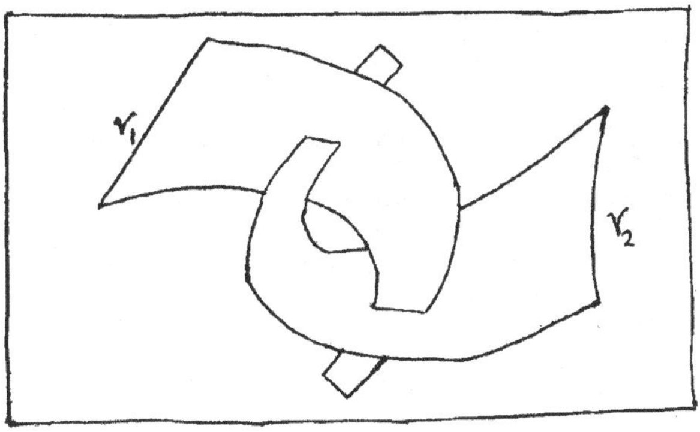

Let $\sigma : V_0 \to V_0$ be a projective transformation of order 2 which permutes P_1 and P_2, and γ_1 and γ_2. σ_0 induces an automorphism $\sigma : U \to U$ which is of order 2.

In this situation, Hironaka has shown (XVIII) that U is a nonprojective variety, and that there is no quotient of U by the action of the group $G = \{1, \sigma\}$ in the category of schemes. (Specifically, there is no geometric quotient in the sense of IV.1.1 below.) Of course, there is a quotient X_a in the category of analytic spaces. In this category, let $\pi_a : U \to X_a$ be the quotient map. π_a is not etale since σ has fixed points. But this is not the source of the problem in finding algebraic structure on X_a. Indeed, if we let $U_1 \subset U$ be the open subspace obtained by deleting all points not laying over γ_1 and γ_2, then U_1 is a projective variety on which $G = \{1,\sigma\}$ acts, and quotients of projective varieties by finite groups always exist. Let X_1 be the open subspace of X_a which is the quotient of U_1 by G. Then X_1 contains all the famification points of $\pi_a : U \to X_a$.

Now let $U_2 \subset U$ be the open subspace on which σ acts freely, and X_2 be the analytic space which is the quotient of U_2 by G. The map $\pi_2 : U_2 \to X_2$ is etale surjective and X_a is obtained by gluing together the two open subsets X_1 and X_2.

Thus we have an example of a finite group G acting freely on a scheme U_2, where the quotient does not exist in the category of schemes.

Let now $R = \coprod_{\sigma \in G} U_2$ be the disjoint union of copies of U_2, one for each $\sigma \in G$. There are two maps, $R \rightrightarrows U_2$. π_1 is the identity map on each $U_2 \subset R$, and π_2 takes u in the σth copy of U_2 to $\sigma(u)$. $R \rightrightarrows U_2$ is then a separated etale equivalence relation whose quotient, if it exists, is identifiable as U_2 modulo the action of G.

In the category of algebraic spaces, quotients of separated etale equivalence relations always exist so our X_2 above has the structure of algebraic space. Hence the original analytic quotient X_a has the structure of a separated algebraic space. (In fact the structure of a nonsingular three-dimensional complex algebraic space.)

As mentioned above, it is known that the quotient of a projective variety, under the action of a finite group, always exists. (I.e., the category of projective varieties is closed under the operation of taking such quotients.) As above, given an arbitrary algebraic space and a fixed-point-free action of a finite group, there is an algebraic space quotient. In fact, P. Deligne has recently shown (unpublished) that the quotient of a separated algebraic space under the action of a finite group always exists in the category of algebraic spaces.

This theorem has more than just theoretical interest. A number of problems in the construction of moduli can be reduced to the problem of finding quotients of schemes by certain algebraic groups (XXVI). Seshadri has shown (XXXI) that the problem of finding a quotient by a large group can in many cases be reduced to finding quotients by finite groups.

Another aspect of the example above is its application to the problem of constructing Hilbert Schemes. If U_2 is a scheme on which a finite group G acts freely, the quotient of

U_2 can be formed as a certain closed subspace C of the Hilbert Scheme H of U_2. Specifically, look at that part of H which parametrizes closed subspaces T of U_2, with T consisting of an n-tuple of distinct points (where n is the order of G). Let $C \subset H$ be those n-tuples which are the orbits of G. If, as above, the quotient of U_2 by G doesn't exist, then neither can H. Hence in general Hilbert Schemes do not exist. For algebraic spaces, however, Artin has proved the following

Theorem ([II],6.2): Let X be an algebraic space of finite type over a field k. Then the functor $Hilb_X(-)$ is represented by an algebraic space locally of finite type over k.

This theorem is proved as an application of Artin's Algebraization Theorem ([II],1.6,1.7) which gives general conditions under which a sheaf on the category of algebraic spaces is representable. We mention another application:

Theorem ([II],7.3): Let $f:X \to S$ be a proper flat morphism of algebraic spaces, where S is locally of finite type over a field or over an excellent Dedekind domain. Suppose f is of finite presentation and cohomologically flat in dimension zero. Then the relative Picard functor $Pic_{X/S}$ (defined in XVII) is represented by an algebraic space locally of finite presentation over S.

<u>Example 4. Modifications (Blowing Up Subspaces). Application to Analytic Spaces.</u>

A <u>modification</u> is a pair consisting of a proper morphism $f:X' \to X$ of algebraic spaces and a closed subspace $Y \subset X$ such that the restriction of f to Y, $f|_{X'-f^{-1}(Y)} : (X'-f^{-1}(Y)) \to (X-Y)$, is an isomorphism.

There are two problems considered in this context. The first is the problem of starting with a given X, and finding a modification $X' \to X$, $Y \subset X$, with Y a proper closed subspace of X, and such that X' is more amenable to algebraic-geometric arguments. The classical result is Chow's lemma, which states that for any scheme X, proper over a noetherian separated base S, a modification $X' \to X$, $Y \subset X$ exists with $X' \to X$ birational, surjective and projective, and with X' projective over S. We prove this below for algebraic spaces, and it is a crucial step in the proof of the Finiteness Theorem IV.4.1, and the Grothendieck Existence Theorem V.6.3. A more recent result is Hironaka's Resolution of Singularities Theorem (XIX), in which X is an integral scheme proper over a field k of characteristic 0, and X' is a nonsingular projective variety.

The converse problem is to determine, given a scheme X' (say, proper over a field k) and a closed subscheme Y' of X', and a proper map $Y' \to Y$, under what conditions there exists a modification $X' \xrightarrow{f} X$, $Y \subset X$, with $f^{-1}(Y) = Y'$ and with the restriction of f to $Y' \to Y$ the given map.

The classical result in this case is the Castelnuovo theorem where X' is a nonsingular surface and Y' is a rational curve on X' with negative self-intersection, and Y is a point. In (III), Artin has proved a theorem showing when such downward modifications can be made. A specific case is the following:

Theorem (III,6.2): Let X' be a nonsingular algebraic space of finite type over a field k and $Y' \subset X'$ a closed subspace. Assume X' and Y' are nonsingular, of dimensions d and d-1 respectively, and that Y' is proper over k. Let $\mathcal{L}$ be the conormal bundle of Y' in X' and assume $\mathcal{L}$ is an ample bundle on Y'. Then there is a modification $f:X' \to X$, $Y \subset X$, with Y a single point and $f^{-1}(Y) = Y'$.

In other words, Y' may be contracted to a point in X. This is analogous to Grauert's Theorem (XVI, Satz 8, p. 353) for analytic spaces. A previous result of this sort was proved by Hironaka (unpublished), who showed that (using our terminology) in some cases, a subspace of a scheme can be contracted in the category of algebraic spaces. This theorem of Hironaka was one of the original motivations for the consideration of algebraic spaces as the natural category in which to do modifications.

Using Artin's theorem, we can construct an example of a singular surface over the complex numbers which is an algebraic space but not a scheme.

Let X_0 be the projective plane over $\mathbf{C}$, and C_0 an elliptic curve in X_0. We assume that C_0 is of degree 3 and is positioned so that the line at infinity cuts C_0 at one of the inflection points of C_0. (E.g., C_0 can be the locus of $y^2z - x^3 + z^2x = 0$.) As is well known (see, e.g., XIII), an elliptic curve has a group structure on its set of points and for our curve C positioned as above, this group structure is given by assuming that for any curve D_0 in the plane, given as the locus of zeros of an algebraic function, if $C_0 \cap D_0 = \{Q_1, Q_2, \ldots, Q_q\}$ (where multiple intersection points appear the appropriate number of times), then $Q_1 + Q_2 + \ldots + Q_q = 0$ in the group.

For any integer n, there are only a finite number of points Q on C_0 whose order in the associated group is n. Since C_0 is uncountable, there is a point P_0 on C_0 of infinite order.

Let X_1 be the surface obtained by blowing up X_0 at P_0. Let C_1 be the proper transform of C_0 and P_1 be the unique point on C_1 over P_0. C_0 in X_0 has self-intersection $C_0 \cdot C_0 = 9$ so C_1 has self-intersection $C_1 \cdot C_1 = 9 - 1 = 8$.

To continue, let X_{i+1} be the surface obtained by blowing up X_i at P_i. Let C_{i+1} be the proper transform of C_i and P_{i+1} the unique point of C_{i+1} over P_i. The self-intersection $C_{i+1} \cdot C_{i+1} = C_i \cdot C_i - 1 = 9 - i$.

Consider X_{10}:

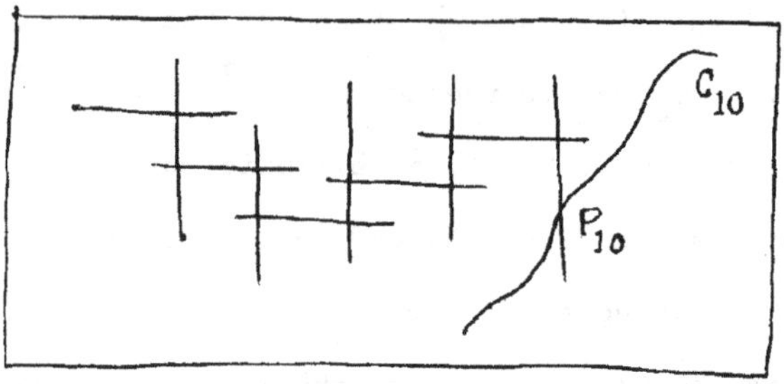

(where the straight lines are the exceptional curves produced in the process). Here $C_{10} \cdot C_{10} = -1$. An interpretation of the self-intersection of C_{10} is that it is the degree of the normal bundle of C_{10} in X_{10}. If we let $\mathcal{L}$ be the dual, the conormal bundle, we have $\deg \mathcal{L} = 1$. Any bundle with positive degree on an elliptic curve is ample. Applying Artin's theorem above, there is a modification $f: X_{10} \to Z$, $z_0 \in Z$, with $f(C_{10}) = z_0$, a point in Z. Z is a singular surface.

We claim Z is not a scheme. For it it were, there would be a curve D on Z with $z_0 \notin D$. $f^{-1}(D) \subset X_{10}$ would be a curve not intersecting C_{10}, and the image D_0 of $f^{-1}(D)$ under the projection $X_{10} \to X_0$ would be a curve in the projective plane intersecting the original elliptic curve C_0 only at the point P_0. Let $mP_0 = D_0 \cap C_0$. Then in the group law on C, $mP_0 = 0$, so P_0 is a torsion point, which is a contradiction.

Another application of Artin's general theorem on modifications is to Moisezon Spaces.

Definition: Let X be a compact complex analytic space (nilpotent elements in the structure sheaf are allowed), and let $C_1,\ldots,C_r$ be the irreducible components of X. Let $d_i = \dim C_i$. Let $K(C_i)$ be the field of meromorphic functions on C_i. The space X will be called a Moisezon Space if the transcendence degree of $K(C_i)$ over $\mathbb{C}$ is d_i for each i.

For a general discussion of these spaces, see Moisezon (XXII,XXIII,XXIV). Note that the transcendence degree of $K(C_i)$ is in any case at most equal to d_i (XXXII).

There is a functor

$$\begin{pmatrix}\text{Algebraic spaces of}\\ \text{finite type over } C\end{pmatrix} \longrightarrow (\text{Complex Analytic Spaces})$$

assigning to each algebraic space its "underlying" analytic space (see I.5.17). By a simple extension of Serre's GAGA (XXX), the restriction of this functor to algebraic spaces proper over C is fully faithful and carries them to compact analytic spaces, the image including at least the projective analytic varieties.

Moisezon Ch. 1, Thm. 1) showed that for a reduced Moisezon space X of dimension n, there is a diagram of irreducible modifications

$$\begin{array}{ccccc} & & X' & & \\ & \stackrel{f}{\swarrow} & & \stackrel{f''}{\searrow} & \\ Y \subset X & & & & X'' \supset Y'' \end{array}$$

where X'' is a projective analytic space, X' is irreducible if X is and of dimension n, and Y, $f^{-1}(Y)$ and $f''^{-1}(Y'')$ are all of dimension less than n.

By GAGA, X'' is the analytic space associated to a projective variety. Artin (III) extends Moisezon's result to the non-reduced case and applies his general theorem to obtain:

<u>Theorem</u> (III,7.3): The above functor induces an equivalence of categories:

(Alg. Spaces proper over C) $\longrightarrow$ (Moisezon Spaces)

In other words, every compact analytic space with enough meromorphic functions "is" an algebraic space in a unique manner.

Having constructed a particular algebraic space, one then can try to show it is a scheme. Some results along this line are proved in this book:

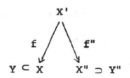

1) If $f:Y \to X$ is locally quasifinite and separated and X is a noetherian scheme, then Y is a scheme. E.g., this applies when f is an immersion. (II.6.16)

2) If $f:Y \to X$ is quasiaffine or quasiprojective and X is a scheme, so is Y. (II.3.8) (II.7.6)

3) If

$$\begin{array}{ccc} Y & \xrightarrow{g} & X \\ & {}_{f}\searrow \quad \swarrow_{h} & \\ & Z & \end{array}$$

is a commutative diagram with Z a noetherian scheme, f affine, g finite surjective and h separated and of finite type, then X is a scheme. (III.4.1)

4) If X is a noetherian separated algebraic space and X_{red} is a scheme, then so is X. (III.3.6)

5) If X is any algebraic space, there is a dense open subspace $U \subset X$ with U a scheme. (II.6.7)

6) With the notation of 5), if X is integral, U contains the generic point. If X is a normal variety, U contains all points of codimension one. (V.4.4)

7) Let $S = \operatorname{Spec} A$, where A is an Artin ring and let X be an algebraic space over S which is an S-algebraic group. Then X is a scheme. (A corollary of 5).

8) Let X be a curve, or a nonsingular surface, over a field k. Then X is a scheme. (V.4.9,10)

We make some comments here on the problems of extending all the results of scheme theory to algebraic spaces. In general, all of the results seem to carry over mutatis mutandis. The exception is when a theorem explicitly uses the fact that around a point p in a scheme X, there is an affine scheme U and a map $i:U \to X$ with $p \in i(U)$ and with i an open immersion, so in particular an injective map. For algebraic spaces, the best one can do is find $U \to X$ etale.

Some of the problems which arise can be easily resolved. For instance, the Grothendieck-topological notion of <u>flask sheaf</u> is not relevant and proofs based on this notion must be modified. This problem shows up here in the sorites on sheaf cohomology of II.4 and the local Ext functor in V.5. The proofs are easily modified.

A more serious problem is that a number of the harder theorems use injectivity of open subsets. For this reason, we have had to construct rather different proofs of the Serre Criterion (III.2.5), Chow's Lemma (IV.3.1) and the Completeness/Extension Lemma (III.1.1). The logical structure of Chapters II-V can be diagrammed:

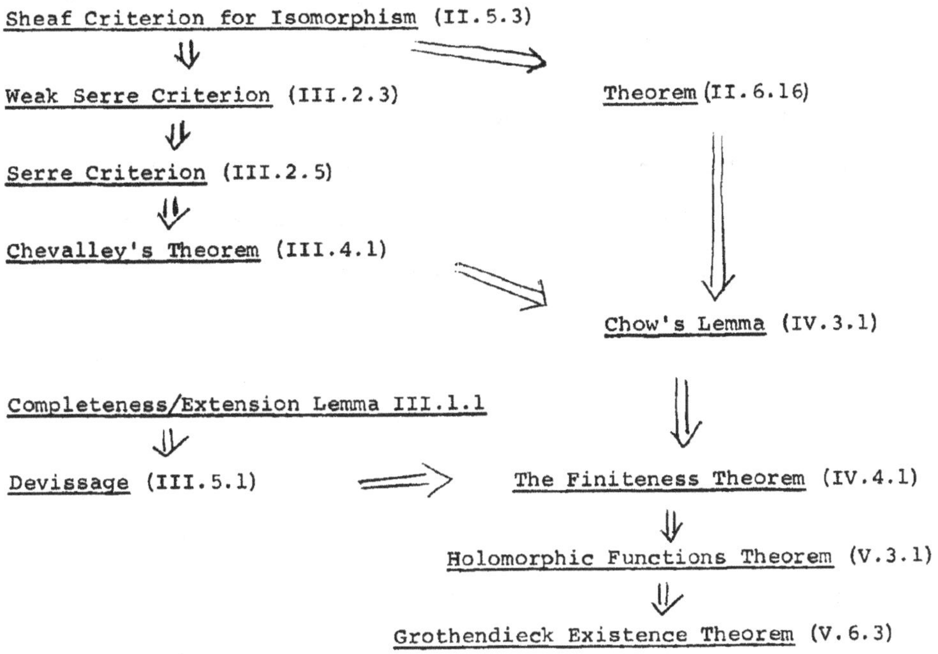

Of these steps, the three mentioned above and II.6.16 have somewhat different proofs than in EGA. The arguments in the other steps are mostly from EGA and in fact most of Chapter V on formal algebraic spaces is practically a straight translation of EGA. (The exception is V.4.4)

Finally, it should be noted that there have been other candidates considered in the search for more general algebraic-geometric objects. These include the notion of Nash manifold (XXVIII) and Matsusaka's notion of Q-variety (XXI). Indeed, in the case of varieties, algebraic spaces are a special case of Q-varieties.

More general objects have been considered, such as Mumford's modular topology (XXVII). We exclude these (for now) since algebraic spaces seem to have more geometric structure (e.g., Artin's theorems). (But see XXXV.)

CHAPTER ONE

THE ETALE TOPOLOGY OF SCHEMES

1. Grothendieck Topologies and Descent Theory............ 29

2. The Zariski Topology of Schemes........................ 38

3. The Flat Topology of Schemes........................... 52

4. The Etale Topology of Schemes.......................... 59

5. Etale Equivalence Relations............................ 72

1. Grothendieck Topologies and Descent Theory

Definition 1.1: A (Grothendieck) Topology τ on a category C consists of a category C = Cat τ and a set Cov τ of families $\{U_i \xrightarrow{\varphi_i} U\}$ $i \in I$ of maps in Cat τ called coverings (where in each covering the range U of the maps φ_i is fixed) satisfying

1) If ϕ is an isomorphism then $\{\phi\} \in$ Cov τ.

2) If $\{U_i \to U\} \in \in$ Cov τ and $\{V_{ij} \to U_i\} \in$ Cov τ for each i then the family $\{V_{ij} \to U\}$ obtained by composition is in Cov τ.

3) If $\{U_i \to U\} \in$ Cov τ and $V \to U \in$ Cat τ is arbitrary then $U_i \times_U V$ exists and $\{U_i \times_U V \to V\} \in$ Cov τ.

<u>Definition 1.2</u>: Let τ be a topology and D a category with products. A <u>presheaf</u> on T with values in D is a functor $F: C^{opp} \to D$. A <u>sheaf</u> is a presheaf satisfying

If $\{U_i \to U\} \in \mathrm{Cov}\,\tau$ then the diagram

$$F(U) \xrightarrow{\pi} \prod_i F(U_i) \underset{\pi_2}{\overset{\pi_1}{\rightrightarrows}} \prod_{i,j} F(U_i \times_U U_j) \text{ is exact.}$$

(Exactness here meaning that π is the difference kernel of π_1, π_2.)

1.3: In general, a Grothendieck topology is not too interesting unless it also satisfies the <u>Axiom A_0</u>: For every object $X \in C$, the contravariant representable functor $\mathrm{Hom}_C(-,X)$ is a sheaf. We write this sheaf $X^{\cdot}$.

Other convenient terminology: For a singleton set $\{f\} \in \mathrm{Cov}\,\tau$ we say f is a <u>covering map</u>, and for a sheaf F of sets, groups, etc. (that is, a sheaf F with values in the category of sets, or the category of groups, or etc.) and an object $X \in C$, we write $\Gamma(X,F) = F(X)$ and refer to the elements of this set (group, etc.) as <u>sections of F on X</u>. Finally, an <u>abelian</u> sheaf is a sheaf of abelian groups.

We assume the reader is somewhat familiar with these notations and with the theories of abelian sheaves and of sheaves of modules over a sheaf of rings. (See, for example, GT and SGAA.)

In the following abstract definitions, it will help to keep in mind the two standard examples:

a) the Global Topology on the category of topological spaces.

Cat τ = the category of topological spaces

Cov τ = all families $\{U_i \xrightarrow{\varphi_i} U\}$ where each map φ_i is an imbedding of U_i as an open subset of U, and where U is covered by the images of the U_i's.

b) the Local Topology on a topological space X.

Cat τ = that category whose objects are open imbeddings $\varphi: U \to X$, and where maps $\sigma: \varphi_1 \to \varphi_2$ are commutative triangles

$$\begin{array}{ccc} U_1 & \xrightarrow{\sigma} & U_2 \\ & {\scriptstyle \varphi_1}\searrow \quad \swarrow{\scriptstyle \varphi_2} & \\ & X & \end{array}$$

Cov τ = all families $\left\{\begin{array}{ccc} U_i & \longrightarrow & U \\ & \searrow \; \swarrow & \\ & X & \end{array}\right\}$ where U is covered by the images of the U_i.

1.4: In the following (through 1.13) let C be a category and $\tau = (C, \mathrm{Cov}\,\tau)$ a topology on C, satisfying the axiom A_0.

<u>Definition 1.5</u>: A class of objects $S \subset C$ is <u>stable</u> (under τ) if for any $\{U_i \to U\} \in \mathrm{Cov}\,\tau$, $U \in S$ if and only if for all i, $U_i \in S$.

<u>Definition 1.6</u>: A <u>closed subcategory</u> D of C is a subcategory D such that

a) D contains all isomorphisms

b) If

$$\begin{array}{ccc} U & \longrightarrow & V \\ {\scriptstyle f'}\downarrow & & \downarrow{\scriptstyle f} \\ X & \longrightarrow & Y \end{array}$$

is a cartesian diagram in C, and $f \in D$, then $f' \in D$.

<u>Definition 1.7</u>: A class D of maps in C is <u>stable</u> (under τ) if D is a closed subcategory and for any $f: X \to Y$ in C, and $\{Y_i \to Y\} \in \mathrm{Cov}\ \tau$, if each $f_i: X \underset{Y}{\times} Y_i \to Y_i \in D$, then $f \in D$.

<u>Definition 1.8</u>: A stable class of maps is <u>local on the domain</u> (under τ) if for any $f: X \to Y \in C$, and any family $\{X_i \xrightarrow{\varphi_i} X\} \in \mathrm{Cov}\ \tau$, $f \in D$ if and only if for all i, $f\varphi_i \in D$.

<u>Definition 1.9</u>: A stable class D of maps of C satisfies <u>effective descent</u> if the following holds: Let $\{U_i \to U\} \in \mathrm{Cov}\,\tau$ and F be a sheaf. Suppose there is a map of sheaves $F \to U^{\cdot}$, such that for each i, the sheaf fiber product $U_i^{\cdot} \underset{U^{\cdot}}{\times} F = W_i^{\cdot}$ for some $W_i \in C$, and suppose the map $W_i \to U_i \in D$. Then it must be that F is representable, say $F = W^{\cdot}$ (and hence the map $W \to U \in D$).

(E.g., in the global standard example 1.4a above, take D to be the set of closed imbeddings.)

<u>Definition 1.10</u>: Let $X \in C$. <u>A cofinal set of coverings</u> $\mathcal{V}$ of X is a set $\{\{X_{ij} \to X\}_{i\in I_j}\}_{j\in J} \subset \mathrm{Cov}\ \tau$, such that for any family $\{U_k \to X\}_{k\in K} \in \mathrm{Cov}\ \tau$, there is a $j \in J$ and a map $n: I_j \to K$, and for each $i \in I_j$, a map $X_{ij} \to U_{n(i)}$ such that $X_{ij} \to U_{n(i)} \to X = X_{ij} \to X$.

<u>Definition 1.11</u>: Let $X \in C$. A <u>local construction</u> Φ on X consists of the following data:

a) A cofinal set $\{\{X_{ij} \to X\}_{i\in I_j}\}_{j\in J}$ of coverings of X.

b) For each $X_{ij} \to X$ appearing in this cofinal set, an object $\Phi(X_{ij}) \in C$, and a map $\Phi(X_{ij}) \to X_{ij}$.

c) For each commutative triangle

$$\begin{array}{ccc} X_{ij} & \longrightarrow & X_{i'j'} \\ & \searrow \quad \swarrow & \\ & X & \end{array}$$

a map $\Phi(X_{ij}) \to \Phi(X_{i'j'})$ such that

$$\begin{array}{ccc} \Phi(X_{ij}) & \longrightarrow & \Phi(X_{i'j'}) \\ \downarrow & & \downarrow \\ X_{ij} & \longrightarrow & X_{i'j'} \end{array}$$

is cartesian.

A local construction Φ on X is <u>effective</u> if there exists an object Y and a map $Y \to X$, such that for each i,j $X_{ij} \times_X Y = \Phi(X_{ij})$. In this case, we write $Y = \Phi(X)$.

Proposition 1.12: Let $X \in C$ and Φ be a local construction on X. Let D be a stable class of maps of C satisfying effective descent and suppose that for each $f_{ij}:X_{ij} \to X$ appearing in a family in the cofinal set associated to Φ, the map $\Phi(X_{ij}) \to X_{ij}$ is in D. Then Φ is effective (and hence the map $\Phi(X) \to X$ is in D).

(To see this, first construct the sheaf which $\Phi(X)$ should represent, and then apply the definition of strict descent.)∎

Definition 1.13: We will say a property P of objects (maps) of C is stable (stable, local on the domain, satisfies effective descent) if the class of all objects (all maps) satisfying P is stable (etc.).

1.14: We now drop our assumption of 1.4 and consider the problem of constructing a topology for a given category C. We assume for convenience that C has fiber products. First we need a preliminary definition.

Definition 1.15: A family $\{U_i \to U\}_{i \in I}$ of maps of C is a universal effectively epimorphic family (UEEF) if for all objects W of C, and maps $W \to U$, and for all objects $V \in C$, the following diagram of sets is exact:

$$\mathrm{Hom}_C(W,V) \to \prod_i \mathrm{Hom}_C(W \times_U U_i, V) \rightrightarrows \prod_{i,j} \mathrm{Hom}_C(W \times_U (U_i \times_U U_j), V)$$

A single map $f:V \to U$ in C is a universal effective epimorphism (UEE) if the family $\{f\}$ is a UEEF.

Definition 1.16: Let B be a closed subcategory of C. The B-topology on C, τ_B (also called the topology on C associated to B) has

$$\text{Cat } \tau_B = C$$

$$\text{Cov } \tau_B = \text{All families } \{U_i \xrightarrow{\varphi_i} U\} \text{ which are UEEF}$$

$$\text{and in which each map } \varphi_i \text{ is in } B.$$

(One can check that this definition satisfies the definition 1.1 and the axiom A_0).

1.17: To get an interesting B-topology, some requirements on B must be satisfied. (For instance, in general B is not stable in the B-topology.) After a preliminary definition, we give a list of possible requirements.

Definition 1.18: An object $\emptyset$ of C is a strict initial object if for all $X \in C$, $\text{Hom}_C(X,\emptyset)$ is empty for X not isomorphic to $\emptyset$, and $\text{Hom}_C(\emptyset,X)$ has exactly one element.

Given a class $\{X_i\}_{i\in I}$ of objects of C, an object X of C is the disjoint union of the class X_i --written $X = \coprod_{i\in I} X_i$-- if X is the categorical sum of the X_i's and for each $i,j \in I$, $X_i \times_X X_j$ is a strict initial object of C. We say C has (finite) disjoint unions if the disjoint union of any (finite) set of objects of C exists.

1.19: We now list some axioms that a closed subcategory B of C might satisfy in order to give a nice topology.

S_1: Let $\{X_i \xrightarrow{\varphi_i} Y\}_{i \in I}$ be a set of maps of C for which the disjoint union $X = \coprod_{i \in I} X_i$ exists, and let $\varphi: X \to Y$ be the induced map. Then $\varphi \in B$ if and only if for all $i \in I$, $\varphi_i \in B$.

(Thus if C has disjoint unions, any covering family $\{U_i \to U\}$ in Cov τ_B can be replaced by a covering map $\coprod_{i \in I} U_i \to U$. The resulting lack of indices often makes arguments much easier.)

S_2: A map $f \in B$ is a universal effective epimorphism if and only if it is an epimorphism.

(Combining S_1, S_2, the B-topology is then just given by the "surjective" maps in B.)

S_3: Let

$$\begin{array}{ccc} X & \xrightarrow{f} & Y \\ & {\scriptstyle h}\searrow \quad \swarrow{\scriptstyle g} & \\ & Z & \end{array}$$

be a commutative diagram in C with $h \in B$.

S_3(a) If $\{f\} \in$ Cov τ_B, then $g \in B$.

S_3(b) If $g \in B$, then $f \in B$.

(If C has disjoint unions, and B satisfies S_1 and S_3(a), then B is stable and local on the domain in the B-topology.)

1.20: The global standard example of 1.4 is obtained by taking B to be the class of all open imbeddings. This class satisfies S_2 and S_3, but B is not local on the domain in the B-topology. The smallest class B' containing B which closed and local on the domain in its own B'-topology is (by definition) the class of local isomorphisms. This class satisfies S_1, S_2 and S_3 and is the prototype of the kind of class of maps one wants when forming topologies. In the category of manifolds, the local isomorphisms are, by the inverse function theorem, exactly those maps satisfying a Jacobian criterion. It is this fact which is exploited to give the definition of etale map, and the etale topology for the category of schemes. (See 4.1, 4.5, 4.6)

The standard local example of 1.4 has B as the class of all maps. This B satisfies S_1 and S_3.

1.21: We insert here, for lack of a better place, a general argument which is used again and again (although often implicitly) in the following.

Lemma 1.21: Let C be a category with fiber products and $D \subset C$ a closed subcategory. Let

$$\begin{array}{ccc} X & \xrightarrow{g} & Y \\ & {\scriptstyle f}\searrow \quad \swarrow{\scriptstyle h} & \\ & Z & \end{array}$$

be a commutative diagram in C. Suppose $f \in D$, and the diagonal map $Y \xrightarrow{\Delta} Y \underset{Z}{\times} Y$ is in D. Then $g \in D$.

<u>Proof</u>. The following two diagrams are cartesian and the composite of the top lines is the map g.

$$\begin{array}{ccc} X & \xrightarrow{1 \times g} & X \underset{Z}{\times} Y \\ {\scriptstyle f}\downarrow & & \downarrow{\scriptstyle f \times 1} \\ Y & \xrightarrow{\Delta} & Y \underset{Z}{\times} Y \end{array} \qquad \begin{array}{ccc} X \underset{Z}{\times} Y & \longrightarrow & Y \\ \downarrow & & \downarrow{\scriptstyle h} \\ X & \xrightarrow[f]{} & Z \end{array}$$

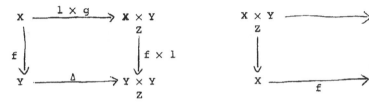

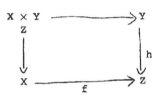

1.22: A final comment. In the following chapters, a lot of proofs will be simplified by assertions that the theorems are local on the objects involved, or local on the domains or ranges of the maps involved. In these cases, what is meant is that all the properties of objects and maps mentioned in the hypotheses and conclusions satisfy the appropriate stability and descent conditions, so that it is sufficient to prove the theorem in some particular case and then invoke the general descent machinery to get the full statement.

2. The Zariski Topology of Schemes

For the sake of completeness we recall the following definitions (with which the reader is assumed to be familiar). More complete treatments can be found in Mumford XXV, Dieudonne IX,X,XI, or the original source, Grothendieck's EGA.

Definition 2.1: Let R be a ring (assumed throughout to be commutative and possess a unit). The prime spectrum of R, Spec R, is the following object:

1) As a set, Spec R is the set of all prime ideals of R. This makes Spec a contravariant functor from the category of rings to that of sets.

2) Spec R is a topological space with the closed sets given by ideals of R. For an ideal I, the corresponding

closed set $V(I)$ is the set of all prime ideals containing I. Spec is thus a functor to the category of topological spaces.

The topology can be defined equivalently as follows: For any element f in R, we write $R_f = \{a/f^n \mid a \in R, n = 0,1,2,\ldots$ modulo the relation $a/f^n = b/f^m$ if and only if there is an integer k with $f^k(f^m a - f^n b) = 0$ in $R\}$. (For details, see Bourbaki, Alg. Comm.)

The natural map $R \to R_f (a \rightsquigarrow a/f^0)$ induces an inclusion Spec $R_f \to$ Spec R, whose image is the complement of the closed set $V((f))$. One can then show that a basis of open sets for the topology on Spec R is given by all open subsets of the form Spec R_f, $f \in R$.

3) The structure sheaf of Spec R is the unique sheaf of rings assigning to every open subset of the form Spec R_f, the ring R_f.

<u>Definition 2.2</u>: A <u>local ringed space</u> $(X, \mathcal{O}_X)$ is a topological space X with a sheaf of rings $\mathcal{O}_X$, such that the stalk of the sheaf at any point $x \in X$ is a local ring (which we denote $\mathcal{O}_{X,x}$). A <u>map</u> $f:(X, \mathcal{O}_X) \to (Y, \mathcal{O}_Y)$ <u>of local ringed spaces</u> is a map of ringed topological spaces such that for every point $x \in X$, the induced map of local rings $\mathcal{O}_{Y,f(x)} \to \mathcal{O}_{X,x}$ maps nonunits to nonunits.

Proposition 2.3: For each ring R, Spec R is a local ringed space and Spec becomes a contravariant functor from the category of rings to the category of local ringed spaces. This functor is full and faithful. ∎

Definition 2.4: The image of this functor (i.e., the dual of the category of rings) is called the category of affine schemes.

We note that the category of affine schemes has fiber products, given by the tensor products of the associated rings, and finite disjoint unions, given by finite products of rings. But observe that the functor Spec does not take infinite products of rings to infinite disjoint unions. (Spec of an infinite product is the Stone-Čech compactification of the union of the Specs.) By extending the category of affine schemes to the category of schemes, and later to the category of algebraic spaces, one gets the correct notion of disjoint union of affine schemes. It will still be true, however, that Spec takes direct limits to inverse limits.

Definition 2.5: An open subscheme Y of an affine scheme X is an open subspace Y of X with the induced sheaf of rings. (Note Y need not itself be an affine scheme.) A closed subscheme Z of an affine scheme X = Spec R is an affine scheme of the form Z = Spec R/I where I is any ideal of R. A subscheme W of X is an open subscheme of a closed subscheme of X. In these cases, the associated inclusion $Y \to X$ is called an open immersion, $Z \to X$ a closed immersion, and $W \to X$ an immersion.

Definition 2.6: A module M over a ring R gives rise to a sheaf of modules $\tilde{M}$ over the structure sheaf on Spec R, by taking $\Gamma(\mathrm{Spec}\ R_f, \tilde{M}) = M \otimes_R R_f$. A general sheaf of modules F on Spec R is called quasicoherent if it is of the form $\tilde{M}$ for some R-module M. If R is noetherian and M is of finite type, we say $\tilde{M}$ is coherent. Similarly for algebras S over R (i.e., maps of rings $R \to S$) we speak of quasicoherent and coherent sheaves of algebras.

Definition 2.7: A scheme is a local ringed space $(X, \mathcal{O}_X)$ such that for every point $p \in X$, there is an open subset $U \subset X$ containing p such that $(U, \mathcal{O}_{X|_U})$ is an affine scheme. (Such a map $U \to X$ with U an affine scheme is called an affine open subscheme of X.)

Definition 2.8: A scheme is quasicompact if it is quasi-compact as a topological space. (We reluctantly follow tradition and write the French "quasicompact" for the English "compact" meaning in either case that every open covering has a finite subcovering.) An affine scheme is automatically quasicompact so for a scheme to be quasicompact it is necessary and sufficient that it have a finite covering by affine schemes.

Definition 2.9: Let $(X, \mathcal{O}_X)$ be a scheme and F an $\mathcal{O}_X$-module. F is quasicoherent if for every point p of X there is an open subset U of X, with $p \in U$, and an exact sequence $\mathcal{O}_X^I \to \mathcal{O}_X^J \to F|_U \to 0$ of $\mathcal{O}_X$-modules (where $\mathcal{O}_X^I$ and $\mathcal{O}_X^J$ denote the sums of the module $\mathcal{O}_X$ over the possibly infinite index sets I and J). When X is an affine scheme, this clearly agrees with the previous definition 2.6.

We say F is locally free if it is quasicoherent and one can choose the open sets U above so that $\mathcal{O}_X^n|_U \cong F|_U$, for some integer n. If F is a locally free sheaf, then for each point p the number n_p is uniquely determined, and is called the rank of F at p. If F has the same rank n_p for every point p, this number is called the rank of F. If n = 1, F is called an invertible sheaf.

Definition 2.10: A map of schemes $f:Y \to X$ is an open immersion (respectively closed immersion, respectively immersion) if for any affine open subscheme U of X, the map $f^{-1}(U) \to U$ is an open immersion (respectively closed immersion, respectively an immersion) in the sense of 2.5. By abuse of language, we say Y is an open subscheme (closed subscheme, subscheme) of X.

If X is a subscheme of Y and Y is a subscheme of Z, then X is a subscheme of Z. The intersection of two subschemes is a subscheme. But note that an arbitrary monomorphism in the category of schemes need not be an immersion.

Proposition 2.11: The category of schemes has fiber products and disjoint sums. The inclusion functor (Affine schemes) $\hookrightarrow$ (Schemes) is full, faithful, and preserves fiber products.

Definition 2.12: A map of schemes $f:X \to Y$ is called injective (surjective, bijective, open, closed) if f is injective (etc.) as a map of topological spaces. One should note that the classes of injective, bijective, open and closed maps of schemes are subcategories but not closed subcategories (in the sense of 1.6). To remedy this, say in the case of open maps, we define a map $f:X \to Y$ to be universally open if for every map of schemes $g:Z \to Y$ the induced map $f':X \times_Y Z \to Z$ is open. The class of universally open maps is then a closed subcategory of (Schemes). Similarly we define universally injective, universally bijective and universally closed. (In E.G.A. the term radiciel is used for universally injective.)

The above definitions are rather inefficient when it comes to proving that a given map f is universally something, since it is apparently necessary to check the topological condition for every possible map $g:Z \to Y$. Luckily one can show (EGA I.3.5) that for the notions of universally injective and bijective, it is sufficient to check when Z is the affine spectrum of an algebraically closed field. For the notions of universally open and closed, the conditions are either more subtle or less general (see EGA II.5.6.3, IV.14).

Definition 2.13: A map of schemes $f:X \to Y$ is separated if the induced diagonal map $\Delta:X \to X \times_Y X$ is a closed immersion. A scheme X is separated if the natural map $X \to \mathrm{Spec}\, Z$ is separated.

An affine scheme is clearly separated. For any scheme X, the diagonal map $X \to X \times X$ is an immersion (EGA I.5.3.9).

Definition 2.14: A map $f:X \to Y$ of schemes is a union of Zariski open sets (Z-open, for short) if X is the disjoint union, $X = \bigsqcup X_i$, of schemes X_i, on each of which the restriction $f|_{X_i}:X_i \to Y$ is an open immersion.

Proposition 2.15: The class of Z-open maps is a closed subcategory of the category of schemes and satisfies axioms S_1, S_2 and S_3(b) (but not S_3(a)). ∎

Definition 2.16: The Zariski topology on the category of schemes is the topology associated with the class of Z-open maps.

Definition-Proposition 2.17: The following properties of schemes are stable in the Zariski topology:

1) X is locally noetherian (i.e., there is a covering of X by affine schemes, each of which is the spectrum

of a noetherian ring. An affine scheme Spec R is locally noetherian if and only if the ring R is noetherian).

2) X is <u>reduced</u> (i.e., there is a covering of X by affine schemes, each of which is the spectrum of a ring with no nilpotent elements).

3) X is <u>normal</u> (i.e., for each point $x \in X$, the local ring $\mathcal{O}_{X,x}$ is an integral domain and is integrally closed in its quotient field).

4) X is <u>nonsingular</u> (i.e., for each point $x \in X$, the local ring $\mathcal{O}_{X,x}$ is a regular local ring. This property is also expressed: X is regular.)

5) X is <u>of dimension n over a ground field k</u> (i.e., there is a map $X \to \operatorname{Spec} k$ of schemes and a covering of X by affine schemes, each of which is of Krull dimension n over k). ∎

<u>Definition-Proposition 2.18</u>: The following properties of maps are stable in the Zariski topology:

1) <u>quasicompact</u> ($f: X \to Y$ is quasicompact if for every quasicompact open subspace $U \subset Y$, $f^{-1}(U) = U \times_Y X$ is quasicompact.)

2) separated

3) quasiseparated ($f:X \to Y$ is quasiseparated if the diagonal map $\Delta:X \to X \times_Y X$ is quasicompact.)

4) universally injective

5) universally bijective

6) universally closed

7) an isomorphism ∎

Definition-Proposition 2.19: The following properties of maps are stable and local on the domain in the Zariski topology:

1) locally of finite type ($f:X \to Y$ is locally of finite type if for any point $x \in X$, there are affine open subschemes $U = \operatorname{Spec} R \subset X$ and $V = \operatorname{Spec} S \subset Y$ with $x \in U$, $f(x) \in V$, $f|_U:U \to V$ and the associated map $S \to R$ of rings makes R finitely generated as an S-algebra.)

2) locally of finite presentation (the same definition as above except we further require R to be finitely presented as an S-algebra.)

3) surjective (note, incidentally, that a surjective map may not be an epimorphism in the category of schemes, and vice-versa. But a Zopen map is surjective if and only if it is an epimorphism.)

4) flat (see def. 3.1)

5) faithfully flat (see def. 3.1)

6) etale (see def. 4.1)

7) universally open

8) locally quasifinite (those maps $f:X \to Y$ which are locally of finite type and for all points $p \to Y$, $f^{-1}(p) = X \times_Y p$ has a discrete underlying topological space.) ∎

<u>Definition-Proposition 2.20</u>: The following classes of maps satisfy effective descent in the Zariski topology:

1) Affine maps ($f:X \to Y$ is <u>affine</u> if for every affine open subscheme U of Y, $f^{-1}(U) = U \times_Y X$ is affine. In this case, if we write $A = f_*\mathcal{O}_X$, A is a quasi-coherent sheaf of $\mathcal{O}_Y$-modules and X and f are uniquely determined by A. We write $X = \underline{\mathrm{Spec}}\ A$. To prove affine maps are effective, one needs to show that for all schemes Z and maps $g:Z \to Y$, $\underline{\mathrm{Spec}}\ (g^*A) = Z \times_Y X$.)

2) Open immersions.

3) Closed immersions

4) Immersions

5) Quasiaffine maps ($f:X \to Y$ is <u>quasiaffine</u> if there exists a scheme W, and maps $g:X \to W$ and $h:W \to Y$ such that $f = hg$, g is a quasicompact open immersion and h is affine. A useful fact--not necessarily for

the proof here--is a theorem of Deligne (EGA IV.18.12.12): Suppose $f:X \to Y$ is separated, quasicompact, locally of finite type, and for every point $y \in Y$, $f^{-1}(y)$ as a topological space is discrete. Then f is quasiaffine.) ∎

Definition-Proposition 2.21: The following are local constructions on a scheme X (where in each case the cofinal set of coverings is all coverings of X by affine schemes):

1) For a given quasicoherent sheaf of $\mathcal{O}_X$-algebras A, the scheme Spec A. (For an inclusion $\phi:U \to X$ with affine U, Spec $\phi(u)$ = Spec $\Gamma(U,A)$.)

2) For any subspace Y of X, the reduced closed subspace $\bar{Y}$ of X, whose set of points is the topological closure of Y.

3) For a given map $f:Y \to X$, with f quasicompact and separated, the scheme-theoretic closure of its image. (By EGA III.1.4.10, or the equivalent proof here (II. 4.6), these conditions imply that $f_* \mathcal{O}_Y$ is a quasicoherent $\mathcal{O}_X$-algebra. Let $\mathcal{I}$ be the sheaf of ideals satisfying the exact sequence $0 \to \mathcal{I} \to \mathcal{O}_X \to f_* \mathcal{O}_Y$. The scheme-theoretic closure of the image of f is defined as Spec $\mathcal{O}_X/\mathcal{I}$. If X is reduced and f is a quasicompact immersion, this gives the previous definition 2).)

4) For a given closed subscheme of X, its open complement.

5) For a given open subscheme of X, its reduced closed complement.

6) The associated reduced scheme of X, X_{red}. (Defined by the requirement that X_{red} is a reduced closed subscheme of X and the map $X_{red} \to X$ is universal for maps of reduced schemes to X.) ∎

2.22: Applying quasicompactness hypotheses, we can also make the following definitions:

Definition A.:

i) A scheme X is noetherian if it is locally noetherian and quasicompact.

ii) A quasicoherent sheaf F on a noetherian scheme is coherent if for any point $x \in X$, there is an open subscheme $U \subset X$, with $x \in U$, finite sets I and J and an exact sequence $\mathcal{O}_X^{\,I} \to \mathcal{O}_X^{\,J} \to F \to 0$. This clearly agrees with our previous definition for a noetherian affine scheme. Note that a locally free sheaf on a noetherian scheme is necessarily coherent.

<u>Definition B.</u>: Let $f:X \to Y$ be a map of schemes.

i) f is <u>of finite type</u> if it is locally of finite type and quasicompact.

ii) f is <u>of finite presentation</u> if it is locally of finite presentation, quasicompact, and the induced map $X \overset{\Delta}{\to} X \times_Y X$ is quasicompact.

iii) f is <u>finite</u> if f is affine, Y is noetherian, and $f_* \mathcal{O}_X$ as an $\mathcal{O}_Y$-module is coherent.

iv) f is <u>quasifinite</u> if f is locally quasifinite and quasicompact. (Deligne's theorem 2.20(5) above now states that a quasifinite separated map is quasi-affine.)

v) f is <u>quasiseparated</u> if the induced map $X \to X \times_Y X$ is quasicompact.

<u>Proposition 2.23</u>: Let X be a quasicompact scheme. Then for any quasiseparated scheme Y (i.e., the map $Y \to Y \times_{(\mathrm{Spec}\ Z)} Y$ is quasicompact) any map $f:X \to Y$ is quasicompact. Also the two projection maps $X \times X \to X$ are quasicompact.

<u>Proof</u>. The main point is that X is quasicompact if and only if the map $X \to \mathrm{Spec}\ Z$ is quasicompact. Each $X \times X \to X$ is a pullback of this map so quasicompact. For the main assertion, use 1.21, the composite map $X \to Y \to \mathrm{Spec}\ Z$, and the quasiseparatedness. ∎

<u>Proposition 2.24</u>: The classes of maps of schemes of finite type, maps of finite presentation, finite maps, quasi-finite maps, and quasiseparated maps are stable in the Zariski topology. ∎

<u>Assumption 2.25</u>: For technical reasons (see II.1.9) we will assume from now on that <u>all the schemes we deal with are quasiseparated</u>. That this is not too serious a restriction can be seen from the following lemma:

<u>Lemma 2.26</u>. Let S be a separated noetherian scheme and U an S-scheme with $U \to S$ locally of finite type. Then U is quasiseparated.

<u>Proof</u>. $U \times U$ must be locally noetherian and $U \to U \times U$ is an immersion. Hence $U \to U \times U$ is quasicompact. ∎

<u>Proposition 2.27</u>: The following is a local construction in the Zariski topology. Let X be a noetherian scheme and F a coherent sheaf on X. For any affine open subscheme Spec R of X, let M be the R-module of finite type for which $\tilde{M} = F|_{\text{Spec } R}$. Let I be the intersection of all maximal ideals m of R for which $mM \neq M$ (equivalently $M \otimes R/m \neq 0$). Put $\Phi(\text{Spec } R) = \text{Spec } R/I$. This local construction is effective

and we write $\Phi(X) = \text{Supp } F$, called the <u>support</u> of F. Supp F is then the reduced closed subspace of X containing all and only those points $x \in X$ for which the stalk of F at x is non-zero. ∎

2.28: In general one deals not with the category of all schemes but rather all schemes over a fixed base scheme S. Here S is taken to be a separated noetherian scheme. An S-scheme is a scheme X with a map $X \to S$ (the structure map). Maps $f: X \to Y$ of S-schemes are required to fit into commutative triangles

$$\begin{array}{ccc} X & \xrightarrow{f} & Y \\ & \searrow \; \swarrow & \\ & S & \end{array} .$$

Since S is quasicompact and separated, an S-scheme X is separated, quasiseparated, quasicompact, etc., in the absolute sense if and only if the structure map $X \to S$ is separated, etc.

Having made these conventions, we will usually suppress all mention of S and just write, e.g., $X \times Y$ for the product $X \underset{S}{\times} Y$ of S-schemes.

3. The Flat Topology of Schemes

We mainly deal here with the flat topology on the category of affine schemes. The generalization to the category of all schemes is indicated in 3.12. The results quoted here constitute the bulk of Bourbaki, Alg. Comm., Ch. 1 (VIII), SGA 60-61, VI,VIII, and EGA IV.2. For an elementary discussion of flat descent, see (VI).

Definition-Proposition 3.1: Let f^*:Spec S → Spec R be a map of affine schemes, or equivalently $f:R \to S$ a map of rings. f is flat if any, hence all, of the following equivalent conditions hold:

1) f_*:(R-modules) → (S-modules) is exact. I.e., for any exact sequence $\mathcal{E}$: $0 \to M' \to M \to M'' \to 0$ of R-modules, the induced sequence $\mathcal{E} \otimes_R S$: $0 \to M' \otimes_R S \to M \otimes_R S \to M'' \otimes_R S \to 0$ is exact.

2) For any exact sequence $\mathcal{E}$ of R-modules of finite type, the induced sequence $\mathcal{E} \otimes_R S$ is exact.

3) For any monomorphism $M' \to M$ of R-modules of finite type, the induced map $S \otimes_R M' \to S \otimes_R M$ is a monomorphism.

4) For any ideal I of R, $I \otimes_R S \to S$ is injective. (In other words, $I \otimes_R S \cong IS$.)

f is faithfully flat (written fflat) if f is flat and any, hence all, of the following equivalent conditions hold:

1') For any sequence $\mathcal{E}$: $0 \to M' \to M \to M'' \to 0$ of R-modules, if the induced sequence $\mathcal{E} \otimes_R S$ is exact, so is $\mathcal{E}$.

2') For any sequence $\mathcal{E}$ of R-modules of finite type, if the sequence $\mathcal{E} \otimes_R S$ is exact, so is $\mathcal{E}$.

3') For any map $M' \to M$ of R-modules, if the induced map $M' \otimes_R S \to M \otimes_R S$ is injective, so is $M' \to M$.

4') For any ideal I of R, $f^{-1}(IS) = I$.

5') For any R-module M, if $M \otimes_R S = 0$, then $M = 0$.

6') For any prime ideal p of R, there is a prime ideal q of S with $f^{-1}(q) = p$. (I.e., f^*:Spec S → Spec R is surjective.)

7') f^*:Spec S → Spec R is a universally effective epimorphism in the category of all schemes. ■

3.2: If $R \to S$ is a faithfully flat map and M is any R-module, then the sequence

$$M \to M \otimes_R S \rightrightarrows M \otimes_R (S \otimes_N S)$$

is exact. Conversely, suppose we are given an S-module N and an $S \otimes_R S$-isomorphism

$$\varphi: N \otimes_R S \to S \otimes_R N$$

which satisfies the "usual cocycle condition": the following diagram of $S \otimes_R S \otimes_R S$ - modules commutes

$$\begin{array}{ccc} N \otimes_R S \otimes_R S & \xrightarrow{\varphi \otimes 1} & S \otimes_R N \otimes_R S \\ & \theta \searrow \quad \swarrow 1 \otimes \varphi & \\ & S \otimes_R S \otimes_R N & \end{array}$$

where θ is "tensoring with φ in the middle"--i.e., $\theta(n \otimes s_1 \otimes s_2) = (1 \otimes s_1 \otimes 1) \cdot d_1(\varphi(n \otimes s_2))$, scalar multiplication in $S \otimes_R S \otimes_R N$, and $d_1(s \otimes n) = s \otimes 1 \otimes n$.

Then there is a unique R-module M with $N = M \otimes_R S$. It is this property, together with the corresponding statement for maps of R-modules, which is the basis of the descent statements in this section.

Proposition 3.3:

a) The class of flat maps of affine schemes is a closed subcategory of (Affine Schemes) and satisfies axioms S_1, S_2 and S_3(a) of 1.19 (but not S_3(b)).

b) An open immersion of affine schemes is flat. ■

Definition 3.4: The *flat topology* on the category of affine schemes is the topology associated with the class of flat maps.

Proposition 3.5: The following properties of maps of affine schemes are stable in the flat topology:

1) Universally injective
2) Universally bijective
3) Universally closed
4) Finite type
5) Finite presentation
6) Finite
7) Etale
8) Quasifinite
9) Isomorphism ■

Proposition 3.6: The following classes of maps are stable and local on the domain in the flat topology on affine schemes:

1) Surjective

2) Flat

3) Fflat

4) Universally open ■

Proposition 3.7: The class of all maps of affine schemes satisfies effective descent in the flat topology. ■

Proposition 3.8: Let $f:X \to Y$ be a map of affine schemes, flat and of finite presentation. Then f is universally open. Furthermore X is noetherian if and only if Y is noetherian. ■

Definition 3.9: A map $f:X \to Y$ of schemes is flat (respectively fflat) if the induced functor

f^*:(Quasicoherent sheaves on Y) $\to$ (Quasicoherent Sheaves on X)

is exact (respectively exact and faithful).

3.10: The topology on the category of schemes determined by the class of flat maps is too fine to be able to generalize the descent theorems 3.3-3.7. The remedy for this situation is indicated by the following proposition:

Proposition 3.10: Let $f:X \to Y$ be faithfully flat and locally of finite presentation. Then there exists a family $\{X_i\}_{i \in I}$ of affine schemes, and for each $i \in I$, a map $X_i \to X$, making X_i a disjoint union of (Zariski) open subschemes of X; a family $\{Y_i\}_{i \in I}$ of affine schemes, indexed by the same set I, and for each $i \in I$, an open immersion $Y_i \to Y$; and finally for each $i \in I$, a map $f_i : X_i \to Y_i$ such that

i) $\{X_i\}$ is a covering of X in the Zariski topology

ii) $\{Y_i\}$ is a covering of Y in the Zariski topology

iii) For each $i \in I$, the following diagram commutes

$$\begin{array}{ccc} X_i & \longrightarrow & X \\ {\scriptstyle f_i}\downarrow & & \downarrow{\scriptstyle f} \\ Y_i & \longrightarrow & Y \end{array}$$

iv) Each f_i is faithfully flat.

Proof. For each $x \in X$, $y = f(x) \in Y$, the condition that f is locally of finite type implies that there are affine open subsets $x \in U'_x \subset X$ and $y \in V_x \subset Y$, with $f(U'_x) \subset V_x$. The restriction of f to U'_x is locally of finite presentation and flat so by 3.8, the image $f(U'_x)$ is open in V_x. Fix U'_x and V_x for the moment. For any point $v \in V_x$, there are affine open sets $U_v \subset X$ and $v \in V_v \subset Y$ with $f(U_v) \subset V_v$, and U_v nonempty. Clearly one can choose $V_v \subset V_x$. Again, each $f(U_v)$ is an open subset of V_v so an open subset of V_x. The set of all such V_v, $v \in V_x$, forms an open covering of V_x. Since V_x is affine, it is quasi-compact so there is a finite set $V_{v_1}, \ldots, V_{v_n}$ such that the union of the images of the corresponding $U_{v_1},\ldots,U_{v_n}$ cover V_x. Let U_x be the disjoint union of U_x', $U_{v_1},\ldots,U_{v_n}$.

Now let the index set I be the collection of points of X, and for each $i \in I$, X_i and Y_i the U_i and V_i constructed above. It is clear from the construction that i-iv are satisfied. ∎

Proposition 3.11: The class of maps of schemes which are flat and locally of finite presentation is a closed subcategory of the category of schemes and satisfies axioms S_1 and S_2. ∎

Definition 3.12: The flat topology on the category of schemes is the topology associated with the set of maps, flat and locally of finite presentation.

One can now prove descent theorems for this topology by proving them for the flat topology on affine schemes and for the Zariski topology on all schemes and then applying 3.10.

Proposition 3.13: In the flat topology on schemes, the following properties of maps are stable: universally injective, universally bijective, universally closed, finite type, finite presentation, finite, etale, quasicompact, separated, quasifinite, and the property of being an isomorphism.

The following properties of maps are stable and local on the domain: surjective, flat fflat, and universally open.

The following properties of maps satisfy effective descent: open immersion, affine, closed immersion, immersion, and quasiaffine.

The following are local constructions in the flat topology: Spec of a quasicoherent sheaf of algebras, the scheme-theoretic closure of the image of a quasicompact separated morphism, the open complement of a closed subscheme.

The following is a stable property of schemes in the flat topology: locally noetherian. ∎

3.14: The reader should take warning that other types of flat topologies have been defined on the category of schemes. Our definition corresponds to the so-called fppf topology (fppf = faithfully flat of finite presentation).

4. The Etale Topology of Schemes

Definition 4.1: A map $f:X \to Y$ of schemes is etale if for any point $x \in X$, there is an affine open subscheme $U = \operatorname{Spec} S \subset X$ with $x \in U$, and an affine open subscheme $V = \operatorname{Spec} R \subset Y$ with $f(x) \in V$, such that $f(U) \subset V$ and the associated map of rings $R \to S$ satisfies the Jacobian Condition: S is an R-algebra of the form $S = R[X_1,\ldots,X_n]/(f_1(X_1,\ldots,X_n),\ldots,f_n(X_1,\ldots,X_n))$ where the determinant of the Jacobian matrix $(\partial f_i/\partial x_j)$ is a unit in S. (The derivations of course are formed as R-derivations in the ring $R[X_1,\ldots,X_n]$).

A consequence of this definition is that a map of rings $R \to S$ is etale if and only if S is an R-algebra of the form $S = R[X_1,\ldots,X_n]/(f_1,\ldots,f_m)$ $n \leq m$, where the ideal in S generated by the $n \times n$ minors of the determinant $\left(\partial f_i/\partial x_j\right)$ is the unit ideal. In particular, by the usual Jacobian criterion for separable algebras over a field, if $R \to S$ is etale and R is a field, then S is a finite product of finite separable field extensions of R.

Alternative Definition 4.2: The class of etale maps is the smallest closed subcategory of the category of schemes which

1) Includes all etale maps $f:X \to Y$ with X, Y affine.

2) Is stable and local on the domain in the Zariski topology and

3) Satisfies axiom S_1 of 1.19.

(Note that we could replace 1 with 1') includes all etale maps of finite type, or with 1") includes all maps of affine schemes satisfying the Jacobian criterion.) ∎

Alternative Definition 4.3: A map $f:X \to Y$ of schemes is unramified if the induced map $X \to X \times_Y X$ is an open immersion. A map of schemes is etale if and only if it is flat, unramified, and locally of finite presentation.

Proof. An etale map is clearly locally of finite presentation. Consider the class of etale maps and the class of maps of schemes which are locally of finite presentation, flat, and unramified. Since these two classes each satisfy 2) and 3) of 4.2, each is determined by the subcategories of just those maps of finite type. By Mumford XXV, p. 436, a map $f:X \to Y$ of finite type is etale if and only if it is flat and the geometric fibers are finite sets of reduced points. This second condition is equivalent to the assertion that the fiber of $f:X \to Y$ over any point $y \in Y$ is a finite disjoint union of affine spectra of finite separable field extensions of the residue field of Y at y. The equivalence of this and the unramified condition is EGA IV.17.4.1. ∎

Alternative Definition 4.4: We say $f:X \to Y$ is formally etale if for every affine scheme Y', and every closed subscheme Y_0' of Y' defined by a nilpotent ideal of $\mathcal{O}_{Y'}$, and every pair of maps g,h making the following commute

$$\begin{array}{ccc} X & \xleftarrow{\ h\ } & Y_0' \\ {\scriptstyle f}\downarrow & & \downarrow{\scriptstyle i} \\ Y & \xleftarrow{\ g\ } & Y' \end{array}$$

there is a unique map $q:Y' \to X$ with $qi = h$ and $fq = g$.

One can then prove (EGA IV.17.6.1, 17.3.1) that a map $f:X \to Y$ of schemes is etale if and only if f is locally of finite presentation and formally etale. ■

<u>Alternative Definition 4.5</u>: Let Y be a locally noetherian scheme. Then a map $f:X \to Y$ is etale if and only if f is locally of finite type and the following condition holds: For every point $x \in X$, write $y = f(x)$, $\mathcal{O}_{X,x}$ the local ring of X at x, $\hat{\mathcal{O}}_{X,x}$ its completion, $\mathcal{O}_{Y,y}$ the local ring of Y at y, k the residue field of $\mathcal{O}_{Y,y}$, $\hat{\mathcal{O}}_{Y,y}$ its completion. Then $\hat{\mathcal{O}}_{X,x}$ is a free $\hat{\mathcal{O}}_{Y,y}$ module and $\hat{\mathcal{O}}_{X,x} \otimes_{\hat{\mathcal{O}}_{Y,y}} k$ is a field, and a finite separable extension of k. (Thus $\hat{\mathcal{O}}_{X,x}$ is a finite $\hat{\mathcal{O}}_{Y,y}$-algebra.) (For a proof, see EGA IV.17.6.3)

In particular, if X and Y are schemes of finite type over a separably closed field k, and $f:X \to Y$ is any k-map, then f is etale if and only if for any closed point $x \in X$, the induced map of complete local rings $\hat{\mathcal{O}}_{Y,f(x)} \to \hat{\mathcal{O}}_{X,x}$ is an isomorphism. (This is clearly a corollary of the above. Another proof can be found in Mumford, XXV, p. 353.) ■

Proposition 4.5a:

1) Etale maps form a closed subcategory of the category of schemes satisfying axioms S_1, S_2, $S_3(a)$ and $S_3(b)$ of 1.19.

2) An etale map is flat and locally of finite presentation. An open immersion is etale.

Proof. Most of this is clear from the above. S_2 follows from the corresponding assertion for the flat topology. $S_3(a)$ and $S_3(b)$ are simple diagram chases, using the alternative definition 4.3. ∎

Definition 4.6: The topology on the category of schemes associated with the class of etale maps is the etale topology.

By 4.5(2), this topology is midway in strength between the Zariski and flat topologies.

Proposition 4.7: Let $f:X \to Y$ be an etale surjective map of schemes. Then there exists a family $\{X_i\}_{i\in I}$ of affine schemes, and for each $i \in I$, a map $X_i \to X$, making X_i a disjoint union of (Zariski) open subschemes of X; a family $\{Y_i\}_{i\in I}$ of affine schemes, indexed by the same set I, and for each $i \in I$, an open immersion $Y_i \to Y$; and finally for each $i \in I$, a map $f_i:X_i \to Y_i$ such that

i) $\{X_i\}$ is a covering of X in the Zariski topology

ii) $\{Y_i\}$ is a covering of Y in the Zariski topology

iii) For each $i \in I$, the following diagram commutes

$$\begin{array}{ccc} X_i & \longrightarrow & X \\ {\scriptstyle f_i}\downarrow & & \downarrow{\scriptstyle f} \\ Y_i & \longrightarrow & Y \end{array}$$

iv) Each f_i is etale surjective and affine (and hence of finite presentation, flat, quasicompact, universally open, and quasifinite.)

Proof. Exactly as in 3.10. ■

4.8: Just as in the flat topology, a proof of a descent theorem for arbitrary etale maps now decomposes into two parts. First we show the theorem for the Zariski topology, and then applying 4.7, we show that the theorem holds etale maps of affine schemes, and hence we are done. To show the theorem for etale maps of affine schemes, it is sufficient to quote the corresponding result (if it is true!) for flat maps of affine schemes. Using this technique, most of the following results are immediate.

Proposition 4.9: The following properties of schemes are stable in the etale topology:

1) locally noetherian

2) reduced

3) normal

4) nonsingular

5) of dimension n over a ground field k

Proof. 1) is proved as indicated in 4.8 above. 2), 3), 4), and 5) are proved in SGA I.9. ■

Proposition 4.10: The following properties of maps of schemes are stable in the etale topology:

1) quasicompact

2) separated

3) universally injective

4) universally closed

5) of finite type

6) of finite presentation

7) finite

8) universally bijective

9) quasifinite

10) being an isomorphism

11) quasiseparated

Proof. All these are stable in both the flat and Zariski topologies. Apply 4.8. ■

Proposition 4.11. The following properties of maps of schemes are stable and local on the domain in the etale topology:

1) surjective

2) flat

3) fflat

4) universally open

5) etale

6) locally of finite presentation

7) locally of finite type

Proof. 1), 2), 3), and 4) follow from the corresponding statements for the flat and Zariski topologies. 5) is part of assertion 4.5(1). As for 6) and 7), see EGA IV.11.3.16. ∎

Proposition 4.12: The following properties of maps of schemes satisfy effective descent in the etale topology:

1) open immersions

2) affine maps

3) closed immersions

4) immersions

5) quasiaffine maps

6) immersions of reduced closed subschemes

Proof. 1) through 5) are effective in the flat and Zariski topologies. 6) follows from 3) and 4.9(2). ∎

Proposition 4.13: The following are local constructions in the etale topology on a scheme X:

1) For a quasicoherent sheaf A of $\mathcal{O}_X$-algebras, $\Phi(X) = \underset{\sim}{\mathrm{Spec}}\, A$.

2) The scheme-theoretic closure of the image of a quasicompact separated map $f: Y \to X$.

3) The open complement of a closed subscheme of X.

4) The reduced closure of a subspace of X.

5) The reduced closed complement of an open subspace of X.

6) X_{red}, the associated reduced scheme.

Proof. 1) through 3) hold in the Zariski and flat topologies. 4), 5), and 6) follow from 4.12(6) above. ∎

4.14: Let X be a scheme and A_Z (respectively A_E) the category of abelian sheaves on X in the Zariski (respectively etale) topology. There is a restriction functor $r: A_E \to A_Z$ which is not in general an isomorphism, even if X is a point, the spectrum of a field. In particular, for an abelian sheaf $F \in A_E$ on X, it can happen that the cohomology $H^q(X,F)$ (defined as the derived functor of $\Gamma(X,-): A_E \to (Ab)$) is nonzero, even when X is a point.

(For a general exposition of the etale cohomology of abelian sheaves on schemes, and proofs of the results above and below, see [GT] and [SGAA]).

For quasicoherent sheaves, the situation is simpler. Let X be a scheme and $\mathcal{O}_X$ its structure sheaf (in the Zariski topology on X). Then $\mathcal{O}_X$ extends uniquely to give a sheaf of rings, which we also denote $\mathcal{O}_X$, on X in the etale topology. A sheaf F of $\mathcal{O}_X$-modules is quasicoherent in the etale topology if there is an etale covering $f:Y \to X$ such that f^*F is the cokernel of a map $\mathcal{O}_Y^{\ I} \to \mathcal{O}_Y^{\ J}$ of free $\mathcal{O}_Y$-modules. A pleasant fact then is that the restriction functor (Quasicoherent sheaves on X in the etale topology) $\to$ (Quasicoherent sheaves on X in the Zariski topology) is an isomorphism. Hence we refer to quasicoherent sheaves on X without specifying the topology.

(The same phenomenon happens in the flat topology--see SGA VIII for details.)

Proposition 4.15: Let X be a scheme, and $H^i(X,-)$ denote the derived functors of the global section functor $\Gamma(X,-)$: (Abelian Sheaves on X in the etale topology) $\to$ (Ab). Let F be a quasicoherent sheaf on X. Then F can be considered as an abelian sheaf on X in the etale topology and we have the following fact: If X is an affine scheme, $H^q(X,F) = 0$, $q > 0$.

Proof. By descent theory (for flat coverings), if $\{U_i \to X\}$ is a covering of X with each U_i affine, the qth Čech cohomology of X with regard to the covering $\{U_i \to X\}$ and sheaf F, vanishes for $q > 0$. In symbols, $\check{H}^q(X,\{U_i\},F) = 0$, $q > 0$. Hence $\check{H}^q(X,F) = 0$. $q > 0$ since such affine etale coverings are cofinal in the class of coverings. This of course holds for any affine scheme U mapping etale to X, $\check{H}^q(U,F) = 0$, $q > 0$.

We now invoke Cartan's Lemma, whose proof in XV.II.5.9.2 generalizes easily to an arbitrary Grothendieck topology. This implies that $H^q(X,F) = 0$. $q > 0$. ∎

Corollary 4.16: Let $f:X \to Y$ be a map of affine schemes and F a quasicoherent sheaf on X. Then F can be considered as an abelian sheaf on X. Let $R^q f_*$ denote the qth derived functor of the functor f_*: (Abelian sheaves on X in the etale topology) → (Abelian Sheaves on Y in the etale topology). Then $R^q f_*(F) = 0$ for $q > 0$.

Proof. $R^q f_*(F)$ is the sheaf associated to the presheaf $\mathcal{R}^q f_*(F)$ where for an etale map $U \to Y$, $\Gamma(U, \mathcal{R}^q f_*(F)) = H^q(U \times_Y X, F)$. For U affine, $U \times_Y X$ is affine so $\Gamma(U, \mathcal{R}^q f_*(F)) = H^q(U \times_Y X, F) = 0$ by the above proposition 4.15 for $q > 0$. Hence $\mathcal{R}^q f_*(F)$ is a presheaf which is zero on every affine scheme U mapping etale to Y. Since every etale covering of Y by schemes is dominated by an etale covering by affine schemes, the associated sheaf $R^q f_*(F)$ is zero, for $q > 0$. ∎

The next lemmas and resulting proposition will be used in sections I.5 and II.6 to show that algebraic spaces always contain open subschemes.

Lemma 4.17. Let $f:Y \to X$ be a quasicompact separated etale map. (So, in particular, f is quasifinite.) For each point $p \in X$, let $n(p)$ be the number of points in the fiber $f^{-1}(p)$. Then n:(points of X) → (Integers) is upper semicontinuous (in the sense that for every integer m, $\{p \mid n(p) \geq m\}$ is an open subspace of X. If for some open subset U of X, n(p) is constant, then the restriction of f to $Y \times_X U \to U$ is a finite etale map.

<u>Proof</u>. See EGA IV.18.2.8. ∎

<u>Lemma 4.18</u>. Let $f:Y \to X$ be a map of finite presentation of schemes and suppose X is an affine scheme. Then there is a map of noetherian schemes $Y_o \to X_o$ and a map $X \to X_o$ such that $Y = X \times_{X_o} Y_o$. If $Y \to X$ is etale quasicompact and separated, $Y_o \to X_o$ can be found etale quasicompact and separated.

<u>Proof</u>. See EGA IV.8.9.1. ∎

<u>Proposition 4.19</u>: Let $f:Y \to X$ be an etale quasicompact separated map. Then there is a dense open subscheme U of X such that the restriction of $f:Y \times_X U \to U$ is a finite etale map. If X is quasicompact, U can be taken to be affine.

<u>Proof</u>. The first assertion is local on X so we can assume X is affine. Consider first the case where X is noetherian and irreducible. Here we apply Lemma 4.17. Let x_o be the generic point of X and $n_o = n(x_o)$, where n(-) is the function of 4.17. Since every nonempty open subset of X contains x_o, $\{p \mid p \in X \text{ and } n(p) \geq n_o + 1\}$ is empty. Hence $U = \{p \mid p \in X, n(p) = n_o\}$ is a dense open subset of X with $U \times_X Y \to U$ finite and etale.

For a general noetherian X, let $X_1,\ldots,X_n$ be the maximal irreducible components of X. Then by the above procedure, open subsets $U_1 \subseteq X_1,\ldots,U_n \subseteq X_n$, $U_i \cap X_j = \emptyset$ if $i \neq j$, can be found satisfying the theorem and we can take $U = U_1 \cup \ldots \cup U_n$.

For the case of general affine schemes X, we can use Lemma 4.18 to find a map of noetherian schemes $Y_o \to X_o$ etale quasicompact and separated. Let $U_o \to X_o$ be a dense open subset with $Y_o \times_{X_o} U_o \to U_o$ finite etale. Let $U = U_o \times_{X_o} X$. Then $U \to X$ is a dense open subset and $Y \times_X U \to U$ is finite etale.

The second statement of the proposition follows from the fact that any quasicompact scheme contains a dense affine open subscheme. ∎

<u>Proposition 4.20</u>: Let X be a scheme and $X_o \subset X$ a closed subscheme defined by a nilpotent ideal of $\mathcal{O}_X$. Consider the functor

(Schemes etale over X) $\to$ (Schemes etale over X_o)

$$Y \rightsquigarrow Y \times_X X_o \quad .$$

This functor is an equivalence of categories.

<u>Proof</u>. See SGA I.5.5, I.8.3. ∎

5. Etale Equivalence Relations

Definition 5.1: Let U be a set and $R \subset U \times U$ an equivalence relation on U. We write π_1 and π_2 for the two projection maps $R \to U \times U \to U$, and say $R \rightrightarrows U$ is a categorical equivalence relation in the category of sets. In a general category C with fiber products, a diagram $R \rightrightarrows U$ in C is called a categorical equivalence relation on U if for all objects $Z \in C$, $\mathrm{Hom}_C(Z,R) \rightrightarrows \mathrm{Hom}_C(Z,U)$ is a categorical equivalence relation in the category of sets.

A map $U \to X$ in C is called the categorical quotient of a categorical equivalence relation $R \rightrightarrows U$ if $U \to X = \mathrm{Cok}(R \rightrightarrows U)$. X is then unique up to unique isomorphism and $U \to X$ is an effective epimorphism. $R \rightrightarrows U$ is effective if it has a categorical quotient $U \to X$ and $R = U \times_X U$.

If $R \rightrightarrows U$ is a categorical equivalence relation in C, and $V \to U$ is any map of C, we define $S = R \times_{(U \times U)} (V \times V)$, and note that $S \rightrightarrows V$ is a categorical equivalence relation, the induced equivalence relation on V.

Proposition 5.2: Let $R \rightrightarrows U$ be a categorical equivalence relation. Then the induced map $\delta : R \to U \times U$ is a monomorphism. Also there is a unique map $i : U \to R$ such that the composite $U \xrightarrow{i} R \xrightarrow{\delta} U \times U$ is the usual diagonal map $\Delta : U \to U \times U$.

<u>Proof</u>. Immediate from the definitions. ∎

<u>Definition 5.3</u>: Let C be a category with fiber products and $\tau = (C, \mathrm{Cov}\,\tau)$ a Grothendieck topology on C satisfying the axiom A_0 of 1.3. A diagram $R \rightrightarrows U$ in C is a <u>τ-equivalence relation</u> if it is a categorical equivalence relation in C and each map π_i is a covering map of τ. Note then that in the category of sheaves of sets on C, $R^{\cdot} \rightrightarrows U^{\cdot}$ is a categorical equivalence relation. A map $U \to X$ of C is a <u>τ-quotient</u> of $R \rightrightarrows U$ if $U^{\cdot} \to X^{\cdot}$ is a categorical quotient of $R^{\cdot} \rightrightarrows U^{\cdot}$. If a τ-quotient exists, it is unique up to unique isomorphism. $R \rightrightarrows U$ is <u>effective</u> (or to be very precise: effective as a τ-equivalence relation) if a τ-quotient exists.

5.4: Let $R \rightrightarrows U$ be a τ-equivalence relation. Consider the associated equivalence relation $R^{\cdot} \rightrightarrows U^{\cdot}$ in the category of sheaves of sets on C. In this category, a categorical quotient $U^{\cdot} \to F$ exists (and so in particular $R^{\cdot} = U^{\cdot} \times U^{\cdot}$). If F is representable, say $F = X^{\cdot}$, i.e., if $R \rightrightarrows U$ is effective as a τ-equivalence relation with τ-quotient $U \to X$, then X is a categorical quotient. But even if F is not representable, $R \rightrightarrows U$ may have a categorical quotient.

The quotient sheaf F of $R^{\cdot} \rightrightarrows U^{\cdot}$ can be constructed as follows: For any object X in C, an element $\gamma \in F(X)$ is given by giving a covering family $\{X_i \to X\}$ of X and elements $\gamma_i \in U^{\cdot}(X_i)$ such that the following condition holds:

$U^{\cdot}(X_i) \to U_{\cdot}(X_i \underset{X}{\times} X_j)$ and $U^{\cdot}(X_j) \to U^{\cdot}(X_i \underset{X}{\times} X_j)$ induce elements $\bar{\gamma}_i, \bar{\gamma}_j$ in $U^{\cdot}(X_i \underset{X}{\times} X_j)$ and we require the pair $<\bar{\gamma}_i, \bar{\gamma}_j>$ to be in $R^{\cdot}(X_i \underset{X}{\times} X_j)$.

Two elements $\alpha, \beta \in F(X)$ are the same if there is a covering family $\{Y_i \to X\}$ where, writing α_i, β_i for the images of α, β in $U(Y_i)$, the pair $<\alpha_i, \beta_i>$ is in $R(Y_i)$.

To see that this defines the quotient sheaf F, let $\mathfrak{F}$ be the presheaf quotient of $R^{\cdot} \overset{\to}{\underset{\to}{}} U^{\cdot}$. Then for all $X \in C$, and covering families $\{X_i \to X\}$, $\mathfrak{F}(X) \to \prod_i \mathfrak{F}(X_i)$ is injective! (Since, for $\alpha, \beta \in \mathfrak{F}(X)$, if α is equivalent to β on each X_i, there is a unique element $<\alpha_i, \beta_i> \in R(X_i)$ for each i, and the sheaf axiom for R gives $<\alpha, \beta> \in R(X)$ so α is equivalent to β on X.)

Thus the quotient sheaf F of $R \overset{\to}{\underset{\to}{}} U$ is $\mathfrak{F}^+$, the usual plus-construction applied to $\mathfrak{F}$ just once. Our construction above just gives this in words in this particular case.

Note also that $\mathfrak{F}(X) \to \prod \mathfrak{F}(X_i)$ injective for every covering family $\{X_i \to X\}$ implies that the natural map $\mathfrak{F}(X) \to F(X)$ is an injective map of sets for all objects X. Hence the sheaf $U \underset{F}{\times} U$ is identical to the presheaf $U \underset{\mathfrak{F}}{\times} U$. R is clearly isomorphic to $U \underset{\mathfrak{F}}{\times} U$ so $R \cong U \underset{F}{\times} U$.

5.5: We now restrict attention to the case where C is a category with fiber products and $\tau = (C, \mathrm{Cov}\ \tau)$ is a B-topology induced by a closed subcategory B of C satisfying S_1, $S_3(a)$ and $S_3(b)$ of 1.19.

The content of 5.4 can be rephrased simply now using the Yoneda Lemma, which asserts that for all X in C, and F a sheaf on $\mathcal{T}$, $F(X) = \mathrm{Hom}_{\mathrm{Sheaves}}(X^{\cdot}, F)$. Thus if F is the quotient of $R^{\cdot} \rightrightarrows U^{\cdot}$ and $\gamma \in F(X)$, there is a covering map $Y \overset{\varphi}{\to} X$ of X and a map $\gamma_1 : Y \to U$ such that the following commutes:

$$\begin{array}{ccccc} Y^{\cdot} \times_X Y^{\cdot} & \overset{\pi_1}{\underset{\pi_2}{\rightrightarrows}} & Y^{\cdot} & \overset{\pi}{\longrightarrow} & X^{\cdot} \\ \Big\downarrow {\scriptstyle \gamma_1 \times_{\gamma} \gamma_2} & & \Big\downarrow {\scriptstyle \gamma_1} & & \Big\downarrow {\scriptstyle \gamma} \\ R^{\cdot} & \overset{\pi_1}{\underset{\pi_2}{\rightrightarrows}} & U^{\cdot} & \underset{\pi}{\longrightarrow} & F \end{array}$$

(where "commuting" for the left hand side of the diagram means that each square $\gamma_1 \pi_i = \pi_i(\gamma_1 \times_{\gamma} \gamma_1)$ commutes.)

Obviously γ is uniquely determined by the maps γ_1 and $\gamma_1 \times_{\gamma} \gamma_1$. Given the two elements $\alpha_1 \beta \in F(X)$, $\alpha = \beta$ iff there is a covering map $\varphi : Y \to X$ and commutative diagrams

$$\begin{array}{ccc} Y^{\cdot} & \overset{\varphi}{\longrightarrow} & X^{\cdot} \\ \Big\downarrow {\scriptstyle \alpha_1} & & \Big\downarrow {\scriptstyle \alpha} \\ U^{\cdot} & \longrightarrow & F \end{array} \qquad\qquad \begin{array}{ccc} Y^{\cdot} & \overset{\varphi}{\longrightarrow} & X^{\cdot} \\ \Big\downarrow {\scriptstyle \beta_1} & & \Big\downarrow {\scriptstyle \beta} \\ U^{\cdot} & \longrightarrow & F \end{array}$$

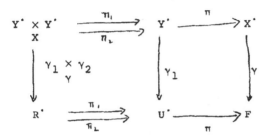

and a map $Y^{\cdot} \to R^{\cdot}$ such that

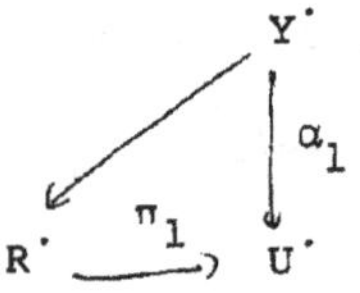

and

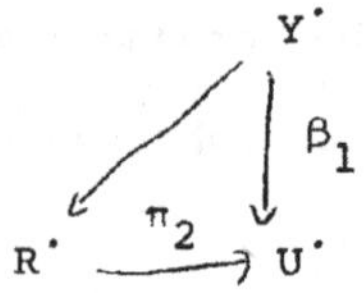

both commute.

<u>Proposition 5.6</u>:

a) If $\pi: U \to X$ is a covering map of the topology τ, then $U \underset{X}{\times} U \rightrightarrows U$ is a τ-equivalence relation with τ-quotient $\pi: U \to X$.

b) Conversely, if $R \rightrightarrows U$ is an effective τ-equivalence relation with τ-quotient $U \to X$, then $U \to X$ is a covering map and $R = U \underset{X}{\times} U$. ∎

<u>Proposition 5.7</u>: Let $R \rightrightarrows U$ be a τ-equivalence relation. Let $V \to U$ be any map in B and $S = R \underset{(U \times U)}{\times} (V \times V)$ the induced equivalence relation on V. Then $S \rightrightarrows V$ is a τ-equivalence relation and there is a "commutative" square

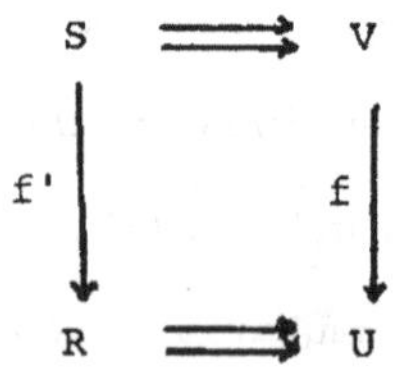

where "commutative" in this situation means that $f\pi_1{}' = \pi_1 f'$, and $f\pi_2' = \pi_2 f'$. Also, if $R \rightrightarrows U$ is effective with τ-quotient $U \to X$. and $S \rightrightarrows V$ is effective with τ-quotient $V \to Y$, the induced map $Y \to X$ is in B. ■

Proposition 5.8: Let D be an effective descent class in C. Let $R \rightrightarrows U$ be an effective τ-equivalence relation with τ-quotient $U \to X$. Suppose there is a "cartesian" diagram in C

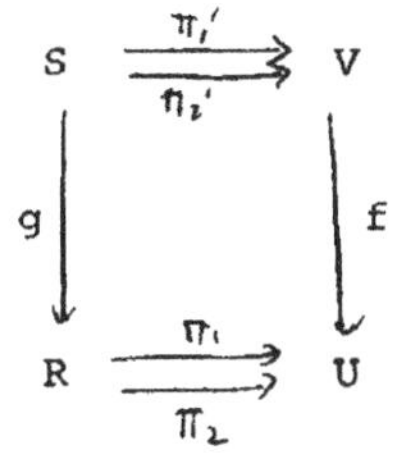

where "cartesian" in this situation means that each square $f\pi_i' = \pi_i g$ is cartesian. Suppose $f \in D$. Then $S \rightrightarrows V$ is an effective τ-equivalence relation. Let $V \to Y$ be the τ-quotient and $h: Y \to X$ the unique map making the following diagram commute:

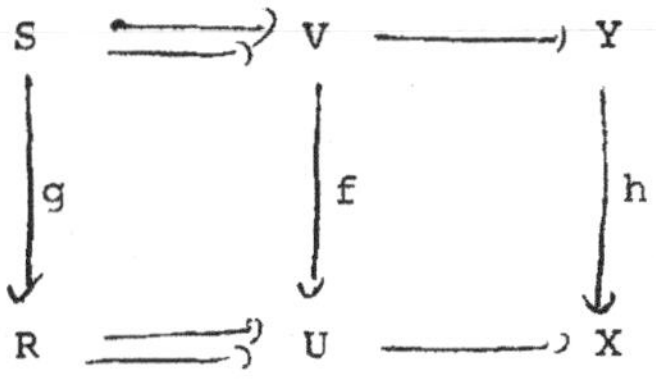

Then $h \in D$ and the square $f\pi' = \pi f$ is cartesian.

<u>Proof</u>. That $S \rightrightarrows V$ is an equivalence relation is a straightforward categorical argument. Since each π_i is a covering map, each π_i' is a covering map so $S \rightrightarrows V$ is a τ-equivalence relation. Let F be the τ-sheaf which is the categorical quotient of $S^{\cdot} \rightrightarrows V^{\cdot}$. Then $V^{\cdot} = U^{\cdot} \times_{X} F$. Applying the notion of effective descent for D, F is representable. Let $Y^{\cdot} = F$. ∎

<u>Proposition 5.9</u>: Let $R \rightrightarrows U$ be a τ-equivalence relation and suppose the induced map $R \to U \times U$ belongs to an effective descent class D in C. Let $U^{\cdot} \to F$ be the categorical quotient of $R^{\cdot} \rightrightarrows U^{\cdot}$ in the category of sheaves and let $V^{\cdot} \to F$ be any map of sheaves with $V \in C$. Then in the cartesian diagram

$$\begin{array}{ccc} V^{\cdot} \times_{F} U^{\cdot} & \longrightarrow & V \\ \downarrow & & \downarrow \\ U^{\cdot} & \longrightarrow & F \end{array}$$

$V^{\cdot} \times_{X} U^{\cdot}$ is represented by an object W in C and the map $W \to V$ is in B, and the map $W \to V \times U$ is in D.

<u>Proof</u>. First consider the special case where $V^{\cdot} \to F$ factors $V^{\cdot} \to U^{\cdot} \to F$. Then $V^{\cdot} \times_{F} U^{\cdot} = V^{\cdot} \times_{U^{\cdot}} U^{\cdot} \times U^{\cdot} = V^{\cdot} \times_{U^{\cdot}} R^{\cdot}$ so is representable. The map $V^{\cdot} \times_{F} U^{\cdot} \to V^{\cdot} \times U^{\cdot}$ is then $V^{\cdot} \underset{U^{\cdot}}{\to} R^{\cdot} \to V^{\cdot} \underset{U^{\cdot}}{\times} (U^{\cdot} \times U^{\cdot})$ so is in D. Also, $V^{\cdot} \times_{F} U^{\cdot} \to V^{\cdot}$ is $V^{\cdot} \times_{U^{\cdot}} R^{\cdot} \to V^{\cdot} = V^{\cdot} \times_{U^{\cdot}} U^{\cdot}$ so is a covering map.

For a general $\gamma : V^{\cdot} \to F$, there is an object W in C and a covering map $W \xrightarrow{\varphi} V$ and a commutative diagram

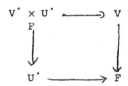

$$\begin{array}{ccccc} W^{\cdot} \underset{V^{\cdot}}{\times} W^{\cdot} & \rightrightarrows & W^{\cdot} & \xrightarrow{\varphi^{\cdot}} & V \\ \downarrow & & \downarrow & & \downarrow \gamma \\ R^{\cdot} & \rightrightarrows & U^{\cdot} & \longrightarrow & F \end{array}$$

(applying 5.5). Consider the cartesian diagram of sheaves:

$$\begin{array}{ccc} W^{\cdot} \underset{F}{\times} W^{\cdot} & \longrightarrow & V^{\cdot} \underset{F}{\times} U^{\cdot} \\ \downarrow & & \downarrow \\ W^{\cdot} \times U^{\cdot} & \longrightarrow & V^{\cdot} \times U^{\cdot} \end{array}$$

By the special case, $W^{\cdot} \times U^{\cdot} \to W^{\cdot} \times U^{\cdot}$ is represented by a map in D. Using the strict descent property of D, the map $V^{\cdot} \underset{F}{\times} U^{\cdot} \to V^{\cdot} \times U^{\cdot}$ is represented by a map in D.

Also,

$$\begin{array}{ccc} W^{\cdot} \underset{F}{\times} U^{\cdot} & \longrightarrow & V^{\cdot} \underset{F}{\times} U^{\cdot} \\ \downarrow & & \downarrow \\ W^{\cdot} & \longrightarrow & V^{\cdot} \end{array}$$

is cartesian so $V^{\cdot} \underset{F}{\times} U^{\cdot} \to V^{\cdot}$ is represented by a covering map. ■

<u>Proposition 5.10</u>: Let $R \rightrightarrows U$ be a τ-equivalence relation where the induced map $R \to U \times U$ belongs to an effective descent class D of C. Let $U^{\cdot} \to F$ be the categorical quotient of $R^{\cdot} \rightrightarrows U^{\cdot}$ in the category of sheaves.

Let $V^{\cdot}_{i} \to F$ and $V^{\cdot}_{2} \to F$ be any two maps of representable sheaves into F. Then $V^{\cdot}_{1} \underset{F}{\times} V^{\cdot}_{2}$ is representable, and the map of C-objects inducing $V^{\cdot}_{1} \times V^{\cdot}_{2} \to V^{\cdot}_{1} \times V^{\cdot}_{2}$ is in D.

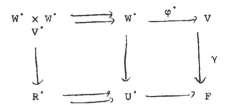

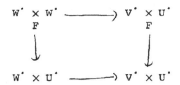

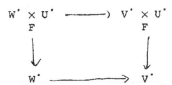

Proof. As in proposition 5.9, one first proves the case where $V^{\cdot}_1 \to F$ and $V^{\cdot}_2 \to F$ factor $V^{\cdot}_1 \to U^{\cdot} \to F$, $V^{\cdot}_2 \to U^{\cdot} \to F$. A similar descent argument then suffices for the general case. ∎

We are going to apply all of this to the etale topology of schemes and define (in II.1.1) the category of algebraic spaces which is in a sense (II.3.14) the closure of the category of schemes under the operation of taking quotients of etale equivalence relations. The examples show that this closure operation is nontrivial--i.e., there are noneffective etale equivalence relations in the category of schemes.

For the rest of this section, we give some facts about etale equivalence relations and some criteria for showing when they are effective.

Proposition 5.11: Let $R \rightrightarrows U$ be an etale equivalence relation (by which phrase we will mean a τ-equivalence relation on the category C, where C is the category of schemes, and τ is the etale topology). Suppose U is a separated scheme. Then the map $R \xrightarrow{\delta} U \times U$ is separated, locally of finite type, and has discrete fibers.

The map $i: U \to R$ is an immersion, identifying U with a component of R (which we call the diagonal part of R).

Proof. Since $R \xrightarrow{\delta} U \times U$ is a monomorphism, it is separated and has discrete fibers. Let $p_i : U \times U \to U$ denote the projection on the ith component, $i = 1,2$. Consider the commutative diagram

$$\begin{array}{ccc} R & \xrightarrow{\delta} & U \times U \\ & {\scriptstyle \pi_i} \searrow \quad \swarrow {\scriptstyle p_i} & \\ & U & \end{array}$$

Since U is separated, p_i is separated. The map $R \xrightarrow{\delta} U \times U$ is separated so π_i is separated.

Also, π_i is locally of finite type since it is etale and, since U is separated, $U \times U \to (U \times U) \underset{U}{\times} (U \times U)$ is a closed immersion also of finite type. Applying 1.21, $R \xrightarrow{\delta} U \times U$ is locally of finite type.

Finally, $i : U \to R$ is a section of the etale map $R \xrightarrow{\pi_i} U$, so by 1.21, i identifies U with a component of R. ∎

Proposition 5.12: Let $R \rightrightarrows U$ be an etale equivalence relation of schemes with U separated. Then the following are equivalent:

1) $R \xrightarrow{\delta} U \times U$ is of finite type

2) $R \xrightarrow{\delta} U \times U$ is quasiaffine

3) $R \xrightarrow{\delta} U \times U$ is quasicompact

4) For all quasicompact schemes V, and open immersions $V \to U$, if $S \rightrightarrows V$ denotes the induced equivalence relation on V, then each π_i' is not only etale but also quasicompact (hence of finite type and quasifinite).

Finally, if R is quasicompact, e.g., if R and U are both schemes of finite type over a ground field, then these equivalent conditions are satisfied automatically. Also, if δ is known to be an immersion, and $U \times U$ is locally noetherian, then the conditions are satisfied.

Proof.

1) <=> 3). $R \xrightarrow{\delta} U \times U$ is locally of finite type by 5.9.

1) <=> 2). By 5.9, $R \xrightarrow{\delta} U \times U$ is separated and has discrete fibers. By Deligne's theorem (2.20(5)), this with the finite type hypothesis implies that $R \xrightarrow{\delta} U \times U$ is quasiaffine. Conversely, quasiaffine implies quasicompact.

3) <=> 4). Suppose $R \xrightarrow{\delta} U \times U$ is quasicompact. Let $V \to U$ be an open immersion with V quasicompact, and $S \rightrightarrows V$ the induced equivalence relation on V which is etale by 5.6. $S \to V \times V$ is quasicompact since it is the pullback of a quasicompact map. Also, V is quasicompact so (by 2.22a) the projections $V \times V \to V$ are quasicompact. Each $\pi_i':S \to V$ is a composite of quasicompact maps, hence quasicompact.

Conversely, suppose condition 4) holds. Let $\{V_i\}$ be the open covering of U by the collection of all quasicompact open subschemes of U. This collection $\{V_i\}$ is closed under the operation of taking finite unions of subschemes. Hence the collection $\{V_i \times V_i\}$ forms an open covering of $U \times U$. To check that $R \xrightarrow{\delta} U \times U$ is quasicompact, it is sufficient to check that for each i, $R \underset{(U\times U)}{\times} (V_i \times V_i) \to (V_i \times V_i)$ is quasicompact since quasicompactness is stable in the Zariski topology. Hence we can assume that U is quasicompact. We are now reduced to showing that if $R \rightrightarrows U$ is an equivalence relation with each π_i etale and quasicompact, then $R \xrightarrow{\delta} U \times U$ is quasicompact. Consider now the commutative triangle

$$\begin{array}{ccc} R & \xrightarrow{\delta} & U \times U \\ & {\scriptstyle \pi_1} \searrow \quad \swarrow {\scriptstyle p_1} & \\ & U & \end{array}$$

where p_1 is the first projection. U is separated, so p_1 is separated, hence $U \times U \to (U \times U) \underset{U}{\times} (U \times U)$ is a closed immersion. A closed immersion is quasicompact. Applying 1.21, we are done; $R \xrightarrow{\delta} U \times U$ must be quasicompact.

Finally, if R is quasicompact and $U \times U$ is separated, applying 1.21 to the composite $R \xrightarrow{\delta} U \times U \to \operatorname{Spec} Z$, we see that $R \to U \times U$ is quasicompact. And any immersion into a locally noetherian scheme must be quasicompact. ∎

If one wishes to view an etale equivalence relation $R \rightrightarrows U$ with quotient $U \to X$ as analogous to a specification of "generators" U and "relations" R, for constructing an "algebra" X, condition 4) above could be viewed as saying "among any finite set of generators, there are only a finitely many relations."

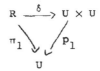

5.13: Let $R \rightrightarrows U$ be an effective etale equivalence relation with quotient $U \xrightarrow{\pi} X$. Then there is a cartesian diagram

$$\begin{array}{ccc} R = U \underset{X}{\times} U & \xrightarrow{\delta} & U \times U \\ \downarrow & & \downarrow \pi \times \pi \\ X & \longrightarrow & X \times X \end{array} .$$

$\pi \times \pi$ is an etale surjective map so for any class D of maps of schemes, stable in the etale topology (e.g., immersions, closed immersions, quasicompact maps, quasiaffine maps), $\delta \in D$ if and only if $\Delta \in D$. Thus separation conditions on X (conditions on the map Δ) are equivalent to conditions on $\delta : R \to U \times U$.

. We have already defined X to be separated if Δ is a closed immersion and X quasiseparated if Δ is quasicompact. Thus the equivalent conditions of 5.12 correspond to quasiseparated schemes. One can also show that for any scheme X, Δ is an immersion (EGA I.5.3.9) so our restriction in 2.26 to quasi-separated schemes entails that for all effective etale equivalence relations $R \rightrightarrows U$, $R \to U \times U$ satisfies the conditions of 5.12.

<u>Proposition 5.14</u>: Let $R \rightrightarrows U$ be an etale equivalence relation where each map π_i is finite and U is affine. Then $R \rightrightarrows U$ is effective. If R and U are affine, the quotient $U \to X$ is given by $X = \mathrm{Spec}(\mathrm{Ker}(\Gamma(U, \mathcal{O}_U) \rightrightarrows \Gamma(R, \mathcal{O}_R)))$.

<u>Proof</u>. XVII.212.5.1 ∎

<u>Proposition 5.15</u>: Let $R \rightrightarrows U$ be an etale equivalence relation satisfying the conditions of 5.12. Suppose U is the affine spectrum of a field K. Then there exists a quotient $U \to X$ with X the affine spectrum of a field L (and hence K is a finite separable field extension of L).

<u>Proof</u>. Using the condition 4) of 5.12, R is quasicompact. Then since U is the spectrum of a field, R must be a finite union of affine spectra of finite separable field extensions of K (applying 4.1). Hence each $\pi_i : R \to U$ is finite so by 5.14, a quotient $\pi : U \to X$ exists. π is an epimorphism (5.1) and etale (5.5). Hence X has only one point and by 4.9, X must be reduced. ∎

<u>Proposition 5.16</u>: (Effectivity is local on U in the Zariski Topology.) Let $R \rightrightarrows U$ be an etale equivalence relation with $R \to U \times U$ quasicompact.

a) Suppose $R \rightrightarrows U$ is effective. Let V be an open subscheme of U and $S \rightrightarrows V$ the induced etale equivalence relation. Then $S \rightrightarrows V$ is effective.

b) Let $\{U_i\}_{i \in I}$ be a Zariski open covering of U and for each $i \in I$, $R_i \rightrightarrows U_i$ the induced equivalence relation. Suppose each $R_i \rightrightarrows U_i$ is effective. Then $R \rightrightarrows U$ is effective.

<u>Proof</u>. a) is clear. To prove b), we first need a

<u>Lemma</u>. Let $W \xrightarrow{\varphi} U$ be an open immersion and $S = R \underset{U \times U}{\times} W \times W$ and suppose the quotient V of $S \rightrightarrows W$ exists:

$$\begin{array}{ccccc} S^{\cdot} & \overset{\pi_1}{\underset{\pi_2}{\rightrightarrows}} & W^{\cdot} & \xrightarrow{\pi} & V^{\cdot} \\ \downarrow & & \downarrow{\scriptstyle\varphi} & & \downarrow \\ R^{\cdot} & \overset{\pi_1}{\underset{\pi_2}{\rightrightarrows}} & U^{\cdot} & \xrightarrow[\pi]{} & F \end{array}$$

Suppose $\pi_2(\pi_1^{-1}(\varphi(W))) = U$. Then $V^{\cdot} \to F$ is an isomorphism.

<u>Proof of Lemma</u>. We must show for all affine schemes X, the natural map $V^{\cdot}(X) \to F(X)$ is an isomorphism. We do this by constructing an inverse map $F(X) \to V^{\cdot}(X)$. Let $\alpha \in F(X) = \mathrm{Hom}_{\text{Sheaves}}(X^{\cdot}, F)$. Let $V^{\cdot}_X = V^{\cdot} \underset{F}{\times} X$, $U^{\cdot}_X = U^{\cdot} \underset{F}{\times} X$, $W^{\cdot}_X = W^{\cdot} \underset{F}{\times} X$, $S^{\cdot}_X = S^{\cdot} \times X$, and $R^{\cdot}_X = R^{\cdot} \times X$. By 5.10, all of these fiber products are representable sheaves. Then there is a diagram of schemes:

$$\begin{array}{ccccc} S_X & \rightrightarrows & W_X & \longrightarrow & V_X \\ \downarrow & & \downarrow{\scriptstyle\varphi} & & \downarrow \\ R_X & \rightrightarrows & U_X & \longrightarrow & X \end{array}$$

which (as a short diagram chase shows) satisfies all the original hypotheses: $R_X \rightrightarrows U_X$ is an etale equivalence relation with quotient $U_X \to X$, $W_X \xrightarrow{\varphi} U_X$ is an open immersion, $S_X \rightrightarrows W_X$ is the induced equivalence relation with quotient $W_X \to V_X$ and $\pi_2(\pi_1^{-1}(\varphi(W_X))) = U_X$. Once we show $V_X \to X$ is an isomorphism, the inverse map (of sheaves) $X^{\cdot} \to V^{\cdot}_X = V^{\cdot} \underset{F}{\times} X^{\cdot}$ gives by projection on the first factor a map $\beta : X^{\cdot} \to V^{\cdot}$, i.e., an element $\beta \in V^{\cdot}(X)$. This $\alpha \rightsquigarrow \beta$ gives the required map $F(X) \to V^{\cdot}(X)$ which is inverse to the natural $V^{\cdot}(X) \to F(X)$.

In short, we can in our original hypotheses for the lemma, assume that F is representable. Then since $W \to V$ is etale surjective and $W \to F$ is etale, $V \to F$ is etale. To show $V \to F$ is one-one, it is sufficient to check that two points of V going to the same point in F must be identical--and this is clear. The special hypothesis $U = \pi_2(\pi_1^{-1}(\varphi(W)))$ makes it clear that $V \to F$ is onto. An etale map which is one-one and onto points is an isomorphism, so we are done. ■ (Lemma)

We now return to the original situation, where $\{U_i \to U\}$ is a covering of U by open subsets, on each of which the induced equivalence relation $R_i \rightrightarrows U_i$ is effective. Let V_i be the quotient of $R_i \rightrightarrows U_i$. Then by the lemma there is an open subscheme $W_i \to U$ with $W^{\cdot}_i = U^{\cdot} \times_F V^{\cdot}_i$ and we can replace U_i with W_i.

$\{V_i \to F\}$ is an effective epimorphic family (since its pullback $\{W_i \to U\}$ is), and for each i,j there is a cartesian diagram

$$\begin{array}{ccc} W^{\cdot}_i \times_{U^{\cdot}} W^{\cdot}_j & \longrightarrow & V^{\cdot}_i \times_F V^{\cdot}_j \\ \downarrow & & \downarrow \\ W_i & \longrightarrow & V_i \end{array}$$

(since $W'_i = V'_i \times_F U'$). $W_i \times_U W_j \to W_i$ is an open immersion so $V'_i \times_F V'_j \to V'_i$ is represented by an open subscheme of V_i. Similarly, $V'_i \times V'_j \to V'_j$ is represented by an open subscheme.

Hence F is the scheme which is obtained by gluing each V_i and V_j along the open subscheme $V_i \times_F V_j$. ∎

5.17: Consider the case of an etale equivalence relation $R \rightrightarrows U$ where each of R and U are locally of finite type over the complex numbers. Then R and U have structures as analytic spaces. We write these associated analytic spaces as $R^h \rightrightarrows U^h$.

Now suppose $R \to U \times U$ is an immersion. Then $R^h \to U^h \times U^h$ is an immersion. In R^h the diagonal Δ (image of $U^h \to R^h$) is a component, so there is an open subset W of $U^h \times U^h$ with $R^h \cap W = \Delta \cap W$.

Let $p \in U^h$ be any point. Then W is an open neighborhood of $(p,p) \in \Delta \subset U^h \times U^h$ which doesn't meet $R^h - \Delta$. Since U^h is Hausdorff, there is an open neighborhood U_p of p in U^h with $U_p \times U_p \subset W$. Thus the restriction of the equivalence relation $R^h \rightrightarrows U^h$ to $U_p \subset U^h$ is the trivial equivalence relation $U_p \rightrightarrows U_p$. In particular, the induced equivalence relation on U_p is effective.

Hence U^h has a covering (in the analytic topology, of course) by open subsets U_p on which the induced equivalence relation is effective. The reasoning in the theorem above is clearly applicable to the case of analytic spaces, so we have shown the following proposition.

<u>Proposition 5.18</u>: Let $R \rightrightarrows U$ be an etale equivalence relation of schemes locally of finite type over the complex numbers, with the map $R \to U \times U$ an immersion. Let $R^h \rightrightarrows U^h$ be the induced equivalence relation in the category of analytic spaces. Then $R^h \rightrightarrows U^h$ is effective in the category of analytic spaces. ∎

<u>Proposition 5.19</u>: Let $R \rightrightarrows U$ be an etale equivalence relation satisfying the equivalent conditions of 5.12. Then there exists a dense open subscheme V of U, such that the induced equivalence relation $S \rightrightarrows V$ is effective.

<u>Proof</u>. Let $u \in U$, V_u an open affine subscheme of U containing u, and $S_u \underset{\pi_{2u}}{\overset{\pi_{1u}}{\rightrightarrows}} V_u$ the induced equivalence relation. Each map π_{iu} is quasifinite (applying our hypothesis of 5.12) so, by proposition 4.19, there is a dense open subscheme V_u' of V_u such that the induced equivalence relation $S_u' \rightrightarrows V_u'$ is finite. By 5.12, $S_u' \rightrightarrows V_u'$ is effective. Now it may be that $u \notin V_u'$, but u is in the closure of V_u'.

Let V be the union of all the V_u'. V is dense in U and, locally on V, the induced equivalence relation is effective. Hence, by 5.16, the induced equivalence relation on V is effective. ∎

<u>Corollary 5.20</u>: Under the hypothesis of 5.19, if U is also quasicompact, V can be chosen so that V and S are both affine (and hence, by 5.14, the quotient of $S \rightrightarrows V$ is affine).

<u>Proof</u>. The same as the proof above, now applying the second statement of 4.19. ∎

CHAPTER TWO

ALGEBRAIC SPACES

1. The Category of Algebraic Spaces 91
2. The Etale Topology of Algebraic Spaces............. 101
3. Descent Theory for Algebraic Spaces................ 106
4. Quasicoherent Sheaves andCohomology................ 113
5. Local Constructions................................ 120
6. Points and the Zariski Topology.................... 129
7. Proper and Projective Morphisms.................... 139
8. Integral Algebraic Spaces 144

1. The Category of Algebraic Spaces

We start with the category of schemes over a given separated noetherian base scheme S (all mention of which will be suppressed) and take this category with the etale topology. As in Chapter I, all schemes are assumed quasi-separated, and for any scheme X, we write $X^{\cdot}$ for the associated representable sheaf.

<u>Definition 1.1</u>: An <u>Algebraic Space</u> A is a functor

$$A:(\text{Schemes})^{\text{opp}} \to (\text{Sets})$$

such that

a) A is a sheaf in the etale topology

b) (Local Representability) There exists a scheme U, and a map of sheaves $U^{\cdot} \to A$ such that for all schemes V, and maps $V^{\cdot} \to A$, the (sheaf) fiber product $U^{\cdot} \times_A V^{\cdot}$ is representable and the map $U^{\cdot} \times_A V^{\cdot} \to V^{\cdot}$ is induced by an etale surjective map of schemes.

c) (Quasiseparatedness) Let $U^{\cdot} \to A$ be as in part b. Then the map of schemes inducing $U^{\cdot} \times_A U^{\cdot} \to U^{\cdot} \times U^{\cdot}$ is quasicompact.

A map $U^{\cdot} \to A$ satisfying b) and c) of this definition will be called a <u>representable etale covering</u> of A. A morphism of algebraic spaces is a natural transformation of functors.

1.2: The functor (Schemes) → (Sheaves of sets on schemes) which is full, faithful, and left exact (by the Yoneda Lemma) factors through the category of algebraic spaces. Hence there is a full faithful left exact imbedding (Schemes) → (Algebraic spaces). For the moment (through Definition 2.5) it will be convenient to maintain a distinction between a scheme X and its associated algebraic space $X^{\cdot}$. After that point we identify the two and just write X for both objects. See 2.6 for details.

Proposition 1.3:

a) Let A be an algebraic space, U a scheme, and $U^{\cdot} \to A$ a representable etale covering. Let R be the scheme representing $U^{\cdot} \times_A U^{\cdot}$. Then $R \rightrightarrows U$ is an etale equivalence relation in the category of schemes and A is the categorical quotient of $R^{\cdot} \rightrightarrows U^{\cdot}$. In particular, if $R \rightrightarrows U$ is effective as an etale equivalence relation in the category of schemes, with quotient $U \to V$, then $A = V^{\cdot}$.

b) Let $R \rightrightarrows U$ be an arbitrary etale equivalence relation in the category of schemes. Suppose the map $R \to U \times U$ is quasicompact. (A requirement which is automatically satisfied if R and U are of finite type over the ubiquitous noetherian separated base scheme S—see I.5.12.) Then there is an algebraic space A, unique up to unique isomorphism, and a map of sheaves $U^{\cdot} \to A$, satisfying part b of definition 1.1, with $R^{\cdot} = U^{\cdot} \times_A U^{\cdot}$.

Proof.

a) Indeed $U^{\cdot} \to A$ is a universally effective epimorphism in the category of sheaves. To see the effectivity, let $U^{\cdot} \xrightarrow{\psi} G$ be any map of sheaves so that the following diagram commutes

$$\begin{array}{ccc} R^{\cdot} \rightrightarrows & U^{\cdot} & \xrightarrow{\pi} A \\ & \downarrow{\scriptstyle \psi} & \\ & G & \end{array}$$

We are required to find a map $\varphi: A \to G$ such that $\varphi\pi = \psi$. To construct a map of sheaves, it is sufficient to give, in a natural fashion for every scheme X, a map $A(X) \xrightarrow{\varphi_X} G(X)$. By the Yoneda Lemma, $A(X) = \mathrm{Hom}_{\mathrm{Sheaves}}(X^\cdot, A)$ and $G(X) = \mathrm{Hom}_{\mathrm{Sheaves}}(X^\cdot, G)$. Let α be a map of sheaves, $\alpha: X^\cdot \to A$. We will construct a map $\varphi_X(\alpha): X^\cdot \to G$. Using part b of definition 1.1, there is a scheme V and an etale surjective map of schemes $V \to X$ so that $V^\cdot = U^\cdot \times_A X^\cdot$. Hence we have the following diagram:

$$\begin{array}{ccccc}
(V \times_X V)^\cdot & \rightrightarrows & V^\cdot & \longrightarrow & X^\cdot \\
\downarrow & & \downarrow & & \downarrow{\scriptstyle \alpha} \\
R^\cdot & \rightrightarrows & U^\cdot & \xrightarrow{\pi} & A \\
 & & {\scriptstyle \psi}\downarrow & & \\
 & & G & &
\end{array}$$

where the two maps $(V \times_X V) \rightrightarrows V^\cdot \to G$ are equal. Now using the sheaf axiom for G, there is a unique map $X^\cdot \xrightarrow{\varphi_X(\alpha)} G$ with $V^\cdot \to X^\cdot \to G = V^\cdot \to G$. This map is an element in $G(X)$ and so we have constructed the required $A(X) \to G(X)$.

The fact that $U^\cdot \to A$ is a universal effective epimorphism is now clear. The first assertion of a) now follows from I.5.5. The second assertion is immediate from the definition of V as the quotient of $R \rightrightarrows U$.

b) This is just a reatatement of I.5.9. ∎

<u>Proposition 1.4</u>: Let A_1 and A_2 be algebraic spaces and $U_1^{\cdot} \to A_1$ and $U_2^{\cdot} \to A_2$ be representable etale coverings. Let g and h be maps such that in the following diagram

$$\begin{array}{ccccc}
(U_1 \underset{A_1}{\times} U_1)^{\cdot} & \overset{\pi_1}{\underset{\pi_2}{\rightrightarrows}} & U_1^{\cdot} & \overset{\pi}{\longrightarrow} & A_1 \\
\Big\downarrow g & & \Big\downarrow h & & \\
(U_2 \underset{A_2}{\times} U_2)^{\cdot} & \overset{\pi_1}{\underset{\pi_2}{\rightrightarrows}} & U_2^{\cdot} & \overset{\pi}{\longrightarrow} & A_2
\end{array}$$

$h\pi_1 = \pi_1 g$ and $h\pi_2 = \pi_2 g$. Then there is a unique map $f:A_1 \to A_2$ with $\pi h = f\pi$. Conversely, every map $f:A_1 \to A_2$ is induced in this way for some choice of U_1, U_2, g, h.

<u>Proof</u>. Of course the converse is the only part requiring proof. Let $f:A_1 \to A_2$ be given and $U_2^{\cdot} \to A_2$ and $V_1^{\cdot} \to A_1$ be representable etale coverings $\mathrm{Hom}_{\mathrm{Sheaves}}(V_1^{\cdot}, A_2) = A_2(V_1)$ so the composite map $\gamma: V_1^{\cdot} \to A_2 \in A_2(V_1)$. A_2 is the quotient sheaf of $(U_2^{\cdot} \underset{A_2}{\times} U_2^{\cdot}) \rightrightarrows U_2^{\cdot}$ so by the construction of quotient sheaves, $\gamma \in A_2(V_1)$ is given by a covering $U_1 \to V_1$ and a section $h \in U_2^{\cdot}(U_1)$ such that the two images of h in $U_2^{\cdot} \underset{A_2}{\times} U_2^{\cdot}(U_1 \underset{V_1}{\times} U_1)$ coincide. This gives $h:U_1^{\cdot} \to U_2^{\cdot}$. Clearly the map $U_1^{\cdot} \to V_1^{\cdot} \to A_1$ is also a representable etale covering of A_1. The existence of the map g follows from the universal mapping property of $U_2^{\cdot} \underset{A_2}{\times} U_2^{\cdot}$. ∎

<u>Proposition 1.5</u>: In the category of algebraic spaces, disjoint sums exist. Also fiber products of algebraic spaces exist--i.e., for any pair of maps $A \to C$ and $B \to C$ of algebraic spaces, the sheaf fiber product $A \times_B C$ is an algebraic space.

<u>Proof</u>. The assertion of disjoint sums is clear.

Given maps of algebraic spaces $A \to C$, $B \to C$, we use proposition 1.4 to find representable etale coverings $X' \to A$, $Y' \to B$, and $Z' \to C$, so that the maps $A \to C$ and $B \to C$ are induced by maps $X \to Z$ and $Y \to Z$. We then have a diagram

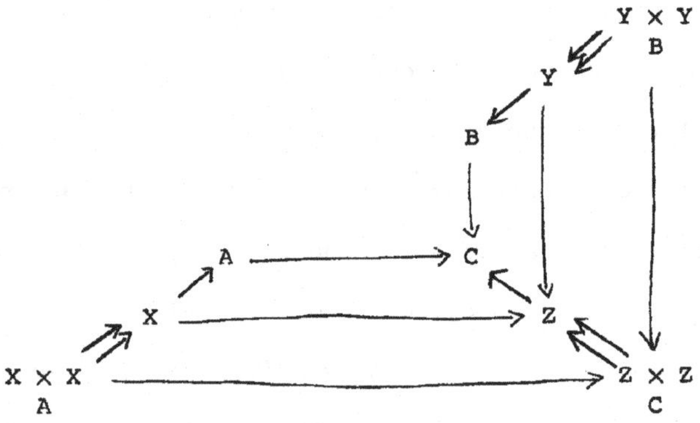

Let F be the sheaf $A \times_C B$. The two maps $X' \times_{Z'} Y' \to X' \to A$ and $X' \times_{Z'} Y' \to Y' \to B$ induce a map $X' \times_{Z'} Y' \to F$. Then

$$(X' \times_{Z'} Y') \times_F (X' \times_{Z'} Y') = (X' \times_A X') \times_{(Z' \times_C Z')} (Y' \times_B Y') .$$

(This is certainly true if $X^{\cdot}$, $Y^{\cdot}$, $Z^{\cdot}$, A, B, C are sets, and to check a statement about inverse limits in an arbitrary category, one can replace each object by its representable functor, and check in the category of sets.) The two maps

$\pi_i : (X^{\cdot} \times_{Z^{\cdot}} Y^{\cdot}) \times_F (X^{\cdot} \times_{Z^{\cdot}} Y^{\cdot}) \to (X^{\cdot} \times_{Z^{\cdot}} Y^{\cdot})$ and

$\pi_i \times \pi_i (X^{\cdot} \times_A X^{\cdot}) \times_{(Z^{\cdot} \times_C Z^{\cdot})} (Y^{\cdot} \times_B Y^{\cdot}) \to X^{\cdot} \times Y^{\cdot}$, are given by

$\pi_i \times \pi_i((x_1,x_2),(y_1,y_2)) = (x_i,y_i)$ and hence are etale. The map $(X^{\cdot} \times_{Z^{\cdot}} Y^{\cdot}) \times_F (X^{\cdot} \times_{Z^{\cdot}} Y^{\cdot}) \to (X^{\cdot} \times_{Z^{\cdot}} Y^{\cdot}) \times (X^{\cdot} \times_{Z^{\cdot}} Y^{\cdot})$ is the map

$(X^{\cdot} \times_A X^{\cdot}) \times_{(Z^{\cdot} \times_C Z^{\cdot})} (Y^{\cdot} \times_B Y^{\cdot}) \to (X^{\cdot} \times X^{\cdot}) \times_{(Z^{\cdot} \times Z^{\cdot})} (Y^{\cdot} \times Y^{\cdot})$

which is quasicompact. Hence we have an etale equivalence relation $(X^{\cdot} \times_{Z^{\cdot}} Y^{\cdot}) \times_F (X^{\cdot} \times_{Z^{\cdot}} Y^{\cdot}) \rightrightarrows (X^{\cdot} \times_{Z^{\cdot}} Y^{\cdot})$. A simple diagram chase shows that the quotient of this equivalence relation is indeed F. Applying 1.3b), F is an algebraic space. ∎

<u>Definition 1.6</u>: Let A be an algebraic space and $U^{\cdot} \to A$ a representable etale covering. Let R be the scheme for which $R^{\cdot} = U^{\cdot} \times U^{\cdot}$. We say that A is <u>locally separated</u> if the map $R \to U \times U$ is a quasicompact immersion. A is <u>separated</u> if $R \to U \times U$ is a closed immersion.

Note for the locally separated case, if U is locally of finite type (over the ubiquitous noetherian separated base scheme S) and $R \to U \times U$ is an immersion, then $R \to U \times U$ is automatically a quasicompact immersion. (See I.2.27)

<u>Proposition 1.7</u>: Let A be an algebraic space and $U^{\cdot} \to A$ a representable etale covering. Let R be the scheme for which $R^{\cdot} = U^{\cdot} \times_A U^{\cdot}$. Let X and Y be schemes and $X^{\cdot} \to A$ and $Y^{\cdot} \to A$ be arbitrary maps of sheaves. Then the sheaf $X^{\cdot} \times_A Y^{\cdot}$ is representable, say by a scheme $W, W^{\cdot} = X^{\cdot} \times_A Y^{\cdot}$, and the map $W \to X \times Y$ is quasicompact.

If furthermore $R \to U \times U$ is an immersion (respectively a closed immersion), then $W \to X \times Y$ is an immersion (respectively a closed immersion).

<u>Proof</u>. We consider two cases.

Case 1. The two maps $X^{\cdot} \to A$ and $Y^{\cdot} \to A$ factor $X^{\cdot} \to U^{\cdot} \to A$ and $Y^{\cdot} \to U^{\cdot} \to A$. Then $X^{\cdot} \times_A Y^{\cdot} = X^{\cdot} \times_{U^{\cdot}} U^{\cdot} \times_A U^{\cdot} \times_{U^{\cdot}} Y^{\cdot} = X^{\cdot} \times_{U^{\cdot}} R^{\cdot} \times_{U^{\cdot}} Y^{\cdot}$ so is representable. The fact that $R \rightrightarrows U$ is an etale equivalence relation and $R \to U \times U$ is quasicompact, implies that $R \to U \times U$ is quasiaffine (see I.5.12). Hence $X^{\cdot} \times_A Y^{\cdot} = X^{\cdot} \times_{U^{\cdot}} R^{\cdot} \times_{U^{\cdot}} Y^{\cdot} \to X^{\cdot} \times_{U^{\cdot}} (U^{\cdot} \times U^{\cdot}) \times_{U^{\cdot}} Y^{\cdot} = X^{\cdot} \times Y^{\cdot}$ is quasiaffine.

Case 2. $X^{\cdot} \to A$ and $Y^{\cdot} \to A$ are arbitrary. Since $U^{\cdot} \to A$ is a representable etale covering, $U^{\cdot} \times_A X^{\cdot}$ and $U^{\cdot} \times_A Y^{\cdot}$ are representable, and the maps representing $U^{\cdot} \times_A X^{\cdot} \to X^{\cdot}$ and $U^{\cdot} \times Y^{\cdot} \to Y^{\cdot}$ are etale surjective. We now apply case one to the maps $U^{\cdot} \times_A X^{\cdot} \to A$ and $U^{\cdot} \times_A Y^{\cdot} \to A$. Consider the cartesian diagram

$$\begin{array}{ccc} (U^{\cdot} \times_A X^{\cdot}) \times_A (U^{\cdot} \times_A Y^{\cdot}) & \longrightarrow & X^{\cdot} \times_A Y^{\cdot} \\ \downarrow & & \downarrow \\ (U^{\cdot} \times_A X^{\cdot}) \times (U^{\cdot} \times_A Y^{\cdot}) & \longrightarrow & X^{\cdot} \times Y^{\cdot} \end{array}$$

The bottom line of this diagram is represented by an etale surjective map of schemes and the left hand side is represented by a quasiaffine map of schemes. By effective descent of quasiaffine maps, the right hand side is represented by a quasiaffine map of schemes. Since a quasiaffine map is quasicompact, we are done. ∎

Corollary 1.8:

a) Let $A:(\text{Schemes})^{\text{opp}} \to (\text{Sets})$ be any sheaf in the etale topology. Let U and V be schemes and $U^{\cdot} \to A$ and $V^{\cdot} \to A$ be maps of sheaves satisfying the local representability condition of Definition 1.1b). Let R and S be the schemes representing $U^{\cdot} \times_A U^{\cdot}$ and $V^{\cdot} \times_A V^{\cdot}$. Then $R \to U \times U$ is quasicompact (respectively a quasicompact immersion, respectively a closed immersion) if and only if $S \to V \times V$ is quasicompact (a quasicompact immersion, a closed immersion). Thus the separation conditions on an algebraic space A are independent of the particular choice of representable etale coverings of A.

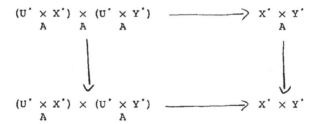

b) Let A be an algebraic space and X a scheme. Let f and g be two elements of A(X)--i.e., f and g are two maps $X^{\cdot} \to A$. Then there exists a scheme Y, and a map of schemes $Y \to X$ such that the induced map $h: Y^{\cdot} \to X^{\cdot}$ is universal for maps of sheaves $h': G \to X^{\cdot}$ satisfying $fh' = gh'$. The map $Y \to X$ is quasicompact. If A is locally separated (respectively separated), then the map $Y \to X$ is a quasicompact immersion (respectively a closed immersion).

Proof. Immediate from the last proposition. ∎

Technical Detail 1.9: The quasiseparatedness assumption in the definition of algebraic space has been used for Proposition 1.3 and hence for the existence of fiber products, a basic construction. The point is that all of the maps $R \to U \times U$ involved are quasiaffine, and the class of quasiaffine maps satisfies effective descent.

Other effective descent classes might have been used. In particular, we could have assumed only that all the maps $R \to U \times U$ involved are immersions (quasicompact or not). Indeed with this assumption it would not be necessary to rule out nonquasicompact schemes.

At this point in the theory a choice of effective descent class had to be made. One effect of this choice is the determination of which pathological examples to allow and disallow.

We've noted (I.5.11) that schemes locally of finite type over a noetherian base scheme S are always quasiseparated and (I.5.12), for any $R \rightrightarrows U$, if $U \times U$ is locally noetherian and $R \to U \times U$ is an immersion, $R \to U \times U$ is quasiaffine. So at the least, the pathological examples involved are all not locally noetherian.

Hence the decision to restrict to the quasiseparated case has been made on other more pragmatic grounds--we have the statements 3.13, 6.2, and 6.7 true for algebraic spaces in general.

2. The Etale Topology of Algebraic Spaces

Definition 2.1: Let $f: A_1 \to A_2$ be a map of algebraic spaces. Using proposition 1.4, choose representable etale coverings $U_1^{\cdot} \to A_1$ and $U_2^{\cdot} \to A_2$ so that f is induced by a map $h: U_1 \to U_2$. We say f is etale if such coverings can be found with h etale. (It is then a simple matter to show that any h' inducing f must be etale.) f is etale surjective if f is etale and a categorical epimorphism. Such an etale surjective f will also be referred to as a covering map.

Proposition-Definition 2.2: For any map of schemes $f:X \to Y$, f is etale (etale surjective) if and only if the associated map of algebraic spaces $f^{\cdot}:X^{\cdot} \to Y^{\cdot}$ is etale (etale surjective) in the above sense. The class of etale maps of algebraic spaces is a closed subcategory of the category of algebraic spaces satisfying axioms S_1, S_2, $S_3(a)$ and $S_3(b)$ of I.1.19. The associated topology on the category of algebraic spaces is the *etale topology* (or sometimes, to be precise, the *global etale topology on the category of algebraic spaces*).

For a particular algebraic space A, the *(local) etale topology* τ *on A* has:

Cat τ = that category whose objects are etale maps $B \to A$, and whose morphisms are commutative triangles

$$\begin{array}{ccc} B_1 & \longrightarrow & B_2 \\ & \searrow \ \swarrow & \\ & A & \end{array}$$

Cov τ = all families $\{B_i \to B\}_{i \in I}$ with $\coprod_{i \in I} B_i \to B$ surjective.

2.3: It is clear from the definitions that a representable etale covering $U^{\cdot} \to A$ is exactly the same as an etale covering $B \to A$, with B a representable sheaf. Conversely, for any covering map $A \to C$ in the etale topology of algebraic spaces, there is a scheme U, and a representable etale covering

$U^{\cdot} \to A$ with the composite $U^{\cdot} \to C$ also a representable etale covering. To carry this one step further, let W be a disjoint union of all the open affine subschemes of U. Then the composite $W^{\cdot} \to C$ is a representable etale covering. Hence we have

<u>Proposition 2.4</u>: In the category of algebraic spaces, any etale covering $A \to B$ of an algebraic space B can be refined to an etale covering $W^{\cdot} \to B$ with W the disjoint union of affine schemes. (Such a covering will be called an <u>affine etale covering</u> of B.)

2.5: As mentioned in 1.2, the functor (Schemes) $\to$ (Algebraic Spaces) taking $X \rightsquigarrow X^{\cdot}$ is full, faithful, and left exact, identifying the category of schemes with a full subcategory of (algebraic spaces). This imbedding is compatible with the etale topologies in both categories.

Furthermore, every sheaf F on the category of schemes in its global etale topology extends uniquely to a sheaf on the category of algebraic spaces in its global etale topology. (By setting, for an algebraic space A with affine etale covering $U^{\cdot} \to A$, and $R^{\cdot} = U^{\cdot} \times_A U^{\cdot}$, $F(A) = \operatorname{Ker}(F(U) \rightrightarrows F(R))$. Of course, conversely, every sheaf on the category of algebraic spaces in the etale topology is determined by its restriction to the category of schemes.

In particular, the category of schemes has the structure sheaf of rings $\mathcal{O}$ (where $\mathcal{O}$(Spec R) = R for any ring R determines $\mathcal{O}$, and this extends to give the structure sheaf of rings on the category of algebraic spaces which we also denote $\mathcal{O}$. Similarly, $\mathcal{O}^*$, the sheaf of units of $\mathcal{O}$, and N, the sheaf of nilpotents of $\mathcal{O}$, extend to algebraic spaces.

For a particular algebraic space A, the restriction of $\mathcal{O}$ to the local etale topology of A is the structure sheaf of A, denoted $\mathcal{O}_A$.

Finally, every algebraic space A is a sheaf on schemes in the etale topology, and as such extends uniquely to a sheaf on the category of algebraic spaces. In other words, every representable functor on the category of algebraic spaces is a sheaf (the axiom A_0).

2.6: The time has come to sort out our notation. From now on, the symbols S,T, U, V, X, ... will denote arbitrary algebraic spaces and the symbols $S^{\cdot}$, $T^{\cdot}$, etc., will denote the associated sheaves on either the category of schemes or algebraic spaces. We consider the category of schemes to be a subcategory of algebraic spaces and say an algebraic space X is a scheme if it lies in this subcategory. Similarly, X is an affine scheme if X lies in the subcategory of affine schemes.

Definition 2.7: An algebraic space X is quasicompact if X has a covering $W \to X$ with W a quasicompact scheme. A map $f:X \to Y$ of algebraic spaces is quasicompact if for every etale map $W \to Y$, with W a quasicompact scheme, $W \times_Y X$ is quasicompact.

X is locally noetherian if there is an etale covering $W \to X$ with W a locally noetherian scheme. X is noetherian if it is quasicompact and locally noetherian.

Proposition 2.8: The class of quasicompact maps of algebraic spaces is stable in the etale topology. ∎

2.9: As mentioned above (2.4) every algebraic space X has an etale covering $W \to X$ with W a disjoint union of affine schemes. If X is quasicompact, we can take W to be a finite disjoint union of affine schemes, hence affine. Thus a quasicompact algebraic space X is the quotient of an etale equivalence relation $R \rightrightarrows W$ with W affine, R a quasicompact open subscheme of an affine sc heme, and, by I.5.12, the maps π_1, π_2 etale and quasicompact, hence quasifinite and of finite type.

If X is quasicompact and separat ed, R and W can both be chosen to be affine. If X is noetherian and separated, R and W can both be chosen to be affine spectra of noetherian rings.

We will later show that every algebraic space X has a covering $\{X_i \xrightarrow{\varphi_i} X\}$, with each X_i a quasicompact algebraic space and each φ_i an "open immersion of algebraic spaces". (Prop. 3.13) Thus any study of the pathologies which "keep algebraic spaces from being schemes" can concentrate on quasicompact algebraic spaces, since the nonquasicompact algebraic spaces are just built in a Zariski-topology fashion from the quasicompact ones.

3. Descent Theory for Algebraic Spaces

We now use descent theory to extend a number of definitions in scheme theory to our context. In each case we implicitly assert that the extended notion is compatible with the original notion when applied to schemes.

Extension 3.1: Let P be a stable property of schemes in the etale topology. Then P extends to a stable property P' of algebraic spaces by taking, for any algebraic space X, with representable etale covering $U \to X$, $P'(X)$ if and only if $P(U)$. Thus we have spoken of locally noetherian algebraic spaces and will speak of reduced algebraic spaces, nonsingular algebraic spaces, normal algebraic spaces, and n-dimensional algebraic spaces over a field k.

Extension 3.2: Let D be a stable class of maps of schemes, local on the domain. Then D extends to a class D' of maps of algebraic spaces which is local on the domain and stable. We define, for any map $f:X \to Y$ of algebraic spaces, $D'(f)$ if and only if for some representable etale covering $U \to Y$ and for some representable etale covering $V \to U \times_Y X$, the induced map of schemes $V \to U$ is in D. Hence we can speak of maps of algebraic spaces being surjective, flat, faithfully flat, universally open, etale, locally of finite presentation, locally of finite type, and locally quasifinite.

Definition 3.3: A map $f:X \to Y$ of algebraic spaces is of finite type if it is locally of finite type and quasicompact. f is of finite presentation if f is locally of finite presentation, quasicompact, and the induced map $X \to X \times X$ is quasicompact. f is quasifinite if f is locally quasifinite and quasicompact.

Proposition 3.4: The classes of maps of algebraic spaces which are of finite type, maps of finite presentation, and quasifinite maps are stable in the etale topology. ∎

Proposition 3.5: Let $f:X \to Y$ be a map of algebraic spaces which is flat and locally of finite presentation. Then f is universally open.

Proof. All three classes of maps involved are defined locally on the domain in the etale topology, so this is a consequence of the corresponding assertion for schemes I.3.8. ∎

3.6: From Chapter III on, we will deal primarily with locally noetherian or noetherian algebraic spaces. As in scheme theory, these categories are not closed under fiber products. But, as in scheme theory, we have the Hilbert Basis Theorem:

Theorem 3.6: Let $f:X \to Y$ be a map of algebraic spaces which is (locally) of finite type. Suppose Y is (locally) noetherian. Then X is (locally) noetherian.

Proof. Exactly as in schemes, we reduce easily to the case where both X and Y are affine schemes and apply the usual Hilbert Basis Theorem. ∎

Corollary 3.7: Let $f:X \to Y$ and $g:Z \to Y$ be maps of (locally) noetherian algebraic spaces. Suppose either f or g is (locally) of finite type. Then $X \times_Y Y$ is (locally) noetherian. ∎

Extension 3.8: Let D be a stable class of maps of schemes satisfying effective descent. Then D extends to a stable effective-descent class D' of maps of algebraic spaces in the following way: A map $f:X \to Y$ of algebraic spaces is in D' if and only if for any scheme Y' and map $Y' \to Y$, $X \times_Y Y'$ is a scheme and $X \times_Y Y' \to Y'$ is in D.

Hence we can talk about open immersions, closed immersions, immersions, affine morphisms, quasiaffine morphisms and reduced *closed immersions of schemes*. We will also use the words *closed subspace*, *open subspace*, and *subspace*. Note in particular that if $f:X \to Y$ is a map of algebraic spaces which is an open immersion, closed immersion, immersion, affine morphism or quasiaffine morphism, and Y is a scheme, then so is X. Thus a subspace of a scheme is a scheme.

Definition 3.9: A map $f:X \to Y$ of algebraic spaces is *quasi-separated*, respectively *locally separated*, respectively *separated* if the induced map $X \to X \times_Y X$ is quasicompact, respectively a quasicompact immersion, respectively a closed immersion.

Proposition 3.10: The classes of quasi-separated, locally separated, and separated maps of algebraic spaces are stable in the etale topology. Also, for any algebraic space X and separated algebraic space Y and any map $f:X \to Y$, X is quasi-quasiseparated, respectively locally separated, respectively separated if and only if the map f is quasiseparated, etc.

Proof. The first statement is clear. For the second, consider first the case where Y is Spec Z, or any separated base schemes over which everything is taken. Let $U \to X$ be a representable etale covering. Then there is a cartesian diagram

$$\begin{array}{ccc} U \times_X U & \longrightarrow & U \times_S U \\ \downarrow & & \downarrow \\ X & \longrightarrow & X \times_S X \end{array}$$

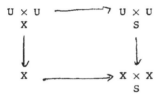

with the right hand side etale surjective. For an arbitrary separated algebraic space Y, apply I.1.21 to the composite $X \to Y \to S$. ∎

<u>Proposition 3.11</u>: Let $f:X \to Y$ be a map of algebraic spaces. Let $g:Y \to X$ be a <u>section</u> of f, i.e., a map satisfying $fg = 1_Y$. If f is separated, g is a closed immersion. If f is etale, g is an open immersion.

<u>Proof</u>. Apply I.1.21 to the composite fg. ∎

<u>Proposition 3.12</u>: Let X be an algebraic space and $\pi:U \to X$ an affine etale covering. Then π is separated. If X is separated, π is an affine map. If X is quasicompact, there is an etale covering $\pi:U \to X$ with U an affine scheme and π etale, separated, and of finite type, hence quasicompact and quasi-finite. ∎

<u>Proposition 3.13</u>: Let X be an algebraic space. Then there is a family of maps $\{X_i \xrightarrow{\varphi_i} X\}_{i\in I}$ with each φ_i an open immersion and each X_i quasicompact, such that the map $\coprod_{i\in I} X_i \to X$ is etale surjective. X is a scheme if and only if each of the X_i is a scheme.

<u>Proof</u>. Let $W \to X$ be an etale covering of X with $W = \coprod_{i\in I} W_i$ the disjoint union of affine schemes W_i. Such coverings exist by 2.4. Let $R = W \underset{X}{\times} W$, and for each $i \in I$, let $R_i = W_i \underset{X}{\times} W_i = R \underset{(W \times W)}{\times} (W_i \times W_i)$. Then $R_i \rightrightarrows W_i$ is

is an etale equivalence relation. By 1.3b, there is an algebraic space X_i with affine etale covering $W_i \to X_i$ and $R_i = W_i \times_{X_i} W_i$. There is an induced map $X_i \to X$ and we claim this map is an open immersion. Let us first sum up in a diagram the given information and name the relevant maps:

$$\begin{array}{ccccc} R_i & \rightrightarrows & W_i & \longrightarrow & X_i \\ \big\downarrow & & \big\downarrow & & \big\downarrow \varphi_i \\ R & \underset{\pi_2}{\overset{\pi_1}{\rightrightarrows}} & W & \xrightarrow{\ \pi\ } & X \end{array}$$

$W \to X$ is a representable etale covering so $W \times_X X_i \to W$ is a map of schemes. A simple diagram chase shows that $X_i \to X$ is a monomorphism. Also, $W_i \to W$ is etale so $X_i \to X$ is etale. Hence $W \times_X X_i \to W$ is an etale monomorphism, hence an open immersion. (Indeed, $W \times_X X_i = \pi_1(\pi_2^{-1}(W_i))$ also showing that $W \times_X X_i \to W$ must be an open immersion.) Hence $X_i \to X$ is an open immersion.

Each X_i is quasicompact since each W_i is quasicompact. The family $\{X_i \xrightarrow{\varphi_i} X\}$ is a covering of X since $\{W_i \to W\}$ is a covering of W. Hence X is the quotient of the etale equivalence relation of algebraic spaces

$\coprod_{i,j\in I} X_i \times_X X_j \rightrightarrows \coprod_{k\in I} X_k$ where each $X_i \times_X X_j$ is an open subspace of X_i and of X_j.

If X is a scheme, every open subspace X_i is a scheme. Conversely, if each of the X_i is a scheme, then X is the scheme obtained by gluing each X_i and X_j together along the common open subsapce $X_i \times_X X_j$. ■

Proposition 3.14: Let $R \overset{\pi_1}{\underset{\pi_2}{\rightrightarrows}} U$ be a categorical equivalence relation of algebraic spaces with each map π_i etale and the induced map $R \to U \times U$ quasiaffine. Then there is a unique algebraic space Z and etale surjective map $U \to Z$ such that $R = U \times_Z U$.

Proof. In the category of sheaves of sets on algebraic spaces, let Z be the quotient sheaf. Then $R^\cdot = U^\cdot \times_Z U^\cdot$. Let $V \to U$ be a representable etale covering of U by a scheme V. Then $V^\cdot \times_Z V^\cdot = V^\cdot \times_{U^\cdot} U^\cdot \times_Z U^\cdot \times_{U^\cdot} V^\cdot = V^\cdot \times_{U^\cdot} R^\cdot \times_{U^\cdot} V^\cdot$ $= R^\cdot \times_{(U^\cdot \times U^\cdot)} (V^\cdot \times V^\cdot)$. Hence $V^\cdot \times_Z V^\cdot \to V^\cdot \times V^\cdot$ is quasi-affine so $V^\cdot \times_Z V^\cdot \rightrightarrows V^\cdot$ is an etale equivalence relation of schemes, whose quotient exists by 1.3b. It is a simple diagram chase to show that this quotient must be Z. ■

4. Quasicoherent Sheaves and Cohomology

Definition 4.1: Let X be an algebraic space and $\mathcal{O}_X$ its structure sheaf of rings. A sheaf of $\mathcal{O}_X$-modules F is *quasi-coherent* if for some covering map $i:U \to X$, with U a scheme, the induced sheaf of $\mathcal{O}_U$-modules, i^*F, is quasicoherent in the usual sense.

For a noetherian algebraic space X, we say F is *coherent* (respectively *locally free of rank r*) if i^*F is coherent (respectively locally free of rank r). A *morphism* of quasi-coherent (coherent) sheaves $F \to G$ is a map of $\mathcal{O}_X$-modules.

Notation 4.2: Let X be an algebraic space. We use the following notation for the following categories:

QCS_X = (Quasicoherent sheaves on X), CS_X = (Coherent Sheaves on X),
MS_X = ($\mathcal{O}_X$-modules), AS_X = (Abelian sheaves on X),
AP_X = (Abelian presheaves on X), and Ab = (Abelian groups).

Proposition 4.3: QCS_X is an abelian category satisfying AB5 and AB3*. The natural maps $QCS_X \to MS_X \to AS_X$ are exact. For noetherian X, the map $CS_X \to QCS_X$ is exact and has the property that for all $0 \to C_1 \to C_2 \to C_3 \to 0$ exact in QCS_X, if two of C_1, C_2, C_3 are in CS_X then the third is also.

Also, if X is quasicompact, then QCS_X has arbitrary sums (= coproducts) where for a family $\{F_i\}$ of quasicoherent sheaves, $\oplus F_i$ can be computed as the sum of the F_i in the category of abelian presheaves on X. (See [GT], p. 49) ∎

4.4: Let $f:X \to Y$. Then $f_*:AS_X \to AS_Y$ has a left adjoint $f^{ab}:AS_Y \to AS_X$. The "same" functor $f_*:MS_X \to MS_Y$ also has a left adjoint $f^m:MS_Y \to MS_X$ where for F an $\mathcal{O}_X$-module, $f^m(F) = f^{ab}(F) \otimes_{f^{ab}\mathcal{O}_Y} \mathcal{O}_X$. Following standard practice we will write f^* for both f^{ab} and f^m but the reader should keep in mind that the two are quite different. (For instance, f^{ab} is exact while f^m is in general only right exact.) Of course if f is etale, then for F an $\mathcal{O}_X$-module, $f^{ab}(F) = f^m(F) = F|_X$.

<u>Proposition 4.5</u>: Let $f:X \to Y$ be a map of algebraic spaces and G a quasicoherent sheaf on Y. Then f^*G (that is, $f^m(G)$) is quasicoherent. f is flat iff f^* takes exact sequences of $\mathcal{O}_Y$-modules into exact sequences of $\mathcal{O}_X$-modules. f is faithfully flat iff f is flat and for any $F \in QCS_Y$, $f^*F = 0$ iff $F = 0$. ∎

<u>Proposition 4.6</u>: Let $f:X \to Y$ be a quasicompact separated map of algebraic spaces and $F \in QCS_X$. Then $f_*F \in QCS_Y$.

<u>Proof</u>. The assertions are local on Y so we can take Y to be affine. Since f is quasicompact, X has an etale covering $W \to X$ with W an affine scheme. Also, since X is separated, $W \times_X W$ is an affine scheme. Thus we have a diagram

$$\begin{array}{ccccc} W \times_X W & \overset{\pi_1}{\underset{\pi_2}{\rightrightarrows}} & W & \xrightarrow{\pi} & X \\ & & & & \downarrow f \\ & & & & Y \end{array}$$

The fact that F is a sheaf on X says that there is an exact sequence

$$F \to \pi_*(F|_W) \overset{\to}{\to} (\pi\,\pi_i)_* F|_{W \underset{X}{\times} W}$$

f_* is left exact so

$$f_*F \to f_*\pi_*F|_W \overset{\to}{\to} (f\,\pi\,\pi_i)_* F|_{W \underset{X}{\times} W}$$

is exact. But the two maps $f\pi$ and $f\pi\pi_i$ are morphisms of affine schemes, so they preserve quasicoherence, and we know that $F|_W$ and $F|_{W \underset{X}{\times} W}$ are quasicoherent. Since $QCS_Y \to MS_Y$ is exact, this proves that f_*F is quasicoherent. ■

<u>Proposition 4.7</u>: Let X be an affine scheme and F a quasi-coherent sheaf on X. Then $H^q(X,F) = 0$, $q > 0$, (where $H^q(X,F)$ is the cohomology of F considered as an abelian sheaf).

Also, if $f:X \to Y$ is a map of affine schemes, and $F \in QCS_X$, then the higher direct images R^qf_*F of F are zero, for $q > 0$, where R^qf_* is here the qth derived functor of the functor $f_*:AS_X \to AS_Y$.

<u>Proof</u>. See I.4.16. ■

<u>Corollary 4.8</u>: Let $f:X \to Y$ be an affine map of algebraic spaces. Then $f_*:QCS_X \to QCS_Y$ is exact. Also, for any $F \in QCS_X$, $R^qf_*(F) = 0$, $q > 0$, where R^qf_* is the qth right derived functor of the functor $f_*:AS_X \to AS_Y$.

Proof. Both assertions are local on Y so we can take Y, hence X, to be affine schemes. The assertion is then clear. ∎

Proposition 4.9: Let E be a separated algebraic space. Then QCS_E has enough injectives and for any I injective in QCS_E, $H^q(E,I) = 0$, $q > 0$, where this is the cohomology of I as an abelian sheaf.

Proof. Let $f:U \to E$ be a covering map, with U a disjoint union of affine schemes, and $F \in QCS_E$. Then $f^*F \in QCS_U$. Let I be an injective envelope of f^*F. (U is a disjoint union of affine schemes so such injective envelopes exist.) Since f_* has an exact left adjoint, f_*I is an injective in QCS_E. Since f_* is left exact, there is a monomorphism $f_*f^*F \to f_*I$. But $F \to f_*f^*F$ is a monomorphism (by the sheaf axiom!) so there is a monomorphism of F into an injective object of QCS_E. Hence QCS_E has enough injectives.

To show that injectives in QCS_E are acyclic for abelian cohomology, it is sufficient to show this for injectives of the form f_*I for I an injective in QCS_U. Indeed, any injective J in QCS_E is a direct summand of such an I, by the argument agove, and cohomology commutes with direct sums. If the abelian cohomology of I vanishes, so must the abelian cohomology of any direct summand J of I.

So let I be injective in QCS_U, for $U \to E$ our covering map. The composite functor $AS_U \to AS_E \to Ab$ gives, for any $F \in AS_U$, a spectral sequence $E_2^{pq} = H^p(E,R^q f_* F) \Rightarrow H^{p+q}(U,F)$. Let now F be our injective sheaf I. By 4.8, $R^q f_* I = 0$ for $q > 0$ and by 4.7, $H^{p+q}(U,I) = 0$, $p+q > 0$. Hence $H^p(E,f_* I) = 0$, $p > 0$. ■

Definition 4.10: Let X be a separated algebraic space. The (quasicoherent) cohomology groups $H^i(X,-)$, $i \geq 0$ of X are the derived functors of the left exact global section functor $\Gamma : QCS_X \to Ab$, $F \rightsquigarrow \Gamma(X,F)$.

Proposition 4.11: For each $i \geq 0$, there is a commutative diagram

$$\begin{array}{ccc} QCS_X & & \\ \downarrow{\scriptstyle \text{natural inclusion}} & \searrow{\scriptstyle H^i(X,-)} & \\ & & Ab \\ & \nearrow{\scriptstyle H^i(X,-)} & \\ AS_X & & \end{array}$$

In other words, the abelian sheaf cohomology coincides with the quasicoherent sheaf cohomology on QCS_X.

Proof. The diagram clearly commutes for $i = 0$, and the natural inclusion is exact. Hence this follows from the usual spectral sequence of a composite functor, which applies in this case since by 4.9, an injective quasicoherent sheaf on X is acyclic for the abelian cohomology. ■

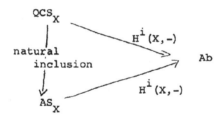

<u>Definition 4.12</u>: Let $f:X \to Y$ be a quasicompact map of separated algebraic spaces. The functor $f_*:QCS_X \to QCS_Y$ is left exact and its higher direct images, $R^q f_*$, $q \geq 0$, are the right derived functors.

<u>Proposition 4.13</u>: (With the assumptions of 4.12) There is a commutative diagram

$$\begin{array}{ccc} QCS_X & \xrightarrow{\quad R^q f_* \quad} & QCS_Y \\ \downarrow & & \downarrow \\ AS_X & \xrightarrow{\quad R^q f_* \quad} & AS_Y \end{array}$$

where the bottom lines are the derived functors of $f_*:AS_X \to AS_Y$. ■

<u>Proposition 4.14</u>: Let $f:X \to Y$ be a quasicompact map of separated algebraic spaces and F a quasicoherent sheaf on X. Then there is a spectral sequence $E_2^{pq} = H^p(Y, R^q f_* F) \Rightarrow H^{p+q}(X,F)$. Let $g:Y \to Z$ be another quasicompact map of algebraic spaces with Z separated. Then there is a spectral sequence $E_2^{pq} = R^p g_*(R^q f_* F) \Rightarrow R^{p+q}(gf)_* F$. ■

<u>Corollary 4.15</u>: Let $f:Y \to X$ be an affine map of algebraic spaces and $F \in QCS_X$. Suppose there is a sheaf $G \in QCS_Y$ such that $f_*G = F$. (In particular, this holds if $F = f_*f^*F$.) Then $H^q(X,F) = H^q(Y,G)$, $q > 0$. ■

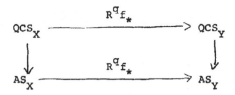

<u>Proposition 4.16</u>: Let X be an algebraic space. The elements of the abelian cohomology group $H^1(X, \mathcal{O}_X^*)$ correspond naturally in a one-one fashion to isomorphism classes of invertible sheaves on X.

<u>Proof</u>. It is a standard fact that in any Grothendieck topology with a final object X, the "twisted forms" of an abelian sheaf F on X are classified by the set $H^1(X, \mathrm{Aut}\, F)$. ∎

<u>Proposition 4.17</u>: Let X be a quasicompact algebraic space, and $\{F_i\}_{i\in I}$ a filtered inductive system of quasicoherent sheaves on X. Then $H^q(X, \varinjlim_I F_i) = \varinjlim_I H^q(X, F_i)$.

<u>Proof</u>. If X is quasicompact, its etale topology is "noetherian" in the sense of [GT], II.5, from which this theorem is quoted. ∎

<u>Proposition 4.18</u>: Let $f: X \to Y$ be a quasicompact separated map of algebraic spaces. Let $\pi: Y' \to Y$ be a flat map of algebraic spaces. Construct a cartesian diagram

$$\begin{array}{ccc} X' & \xrightarrow{\pi'} & X \\ \downarrow{\scriptstyle f'} & & \downarrow{\scriptstyle f} \\ Y' & \xrightarrow{\pi} & Y \end{array}$$

Then $\pi^* f_* = f_*' \pi'^* : QCS_X \to QCS_{Y'}$. (I.e., the natural transformation $f_*'\pi'^* \to \pi^* f_*$ is a natural equivalence.) Indeed, for all $q \geq 0$, $\pi^* R^q f_* = R^q f'_* \pi'^* : QCS_X \to QCS_{Y'}$.

Proof. The assertion is local on Y so we can take Y to be an affine scheme. It is local on Y' so we can take Y' to be an affine scheme. Using the technique of the proof of 4.6, it is local on X so we can take X to be an affine scheme. In the category of affine schemes, the assertion is clear.∎

5. Local Constructions

We now apply the technique of local constructions, as outlined in I.1, I.4, for a number of constructions, always using the class of quasiaffine maps for the strict descent involved and coverings by affine schemes (or disjoint unions of affine schemes) for the cofinal set of coverings involved.

Construction 5.1: Spec. Let X be an algebraic space and A a quasicoherent sheaf of $\mathcal{O}_X$-algebras. For any covering map $\pi: Y \to X$, let $\phi(Y) = \operatorname{Spec} \pi^* A$. By E6A II.1.5.2 this is a local construction. The induced $\phi(X)$ is denoted Spec A.

Note that the map Spec A → X is an affine morphism and any affine morphism $f: Y \to X$ is of the form $Y = \operatorname{Spec} A \to X = \operatorname{Spec} \mathcal{O}_X$ where $A = f_* \mathcal{O}_Y$. (Throughout here, all assertions are local on X, and clear for X affine.)

In particular, if $\mathcal{I} \subset \mathcal{O}_X$ is a sheaf of ideals on X (i.e., a quasicoherent subsheaf of $\mathcal{O}_X$) then $\operatorname{Spec} \mathcal{O}_{X/\mathcal{I}}$ is a closed subspace of X, and every closed subspace of X is so describable.

Also, for $A = \mathcal{O}_X[T_1,\ldots,T_n]$, the polynomial algebra, we have $\underset{\sim}{\mathrm{Spec}}\, A = \mathbb{A}_X^n$, "affine n-space over X".

Proposition 5.2: Let $f:Y \to X$ be a map of algebraic spaces and A a quasicoherent $\mathcal{O}_X$-algebra. Let $X' = \underset{\sim}{\mathrm{Spec}}\, A$ and $Y' = \underset{\sim}{\mathrm{Spec}}\, f^*A$. Then the natural maps $Y' \to X'$, $Y' \to Y$ and $X' \to X$ give a cartesian diagram

$$\begin{array}{ccc} Y' & \longrightarrow & X' \\ \downarrow & & \downarrow \\ Y & \longrightarrow & X \end{array}$$

Proposition 5.3: (The Sheaf Criterion for Isomorphism) Let $f:X \to Y$ be a quasicompact separated map of algebraic spaces with Y an affine scheme. Suppose f is faithfully flat. Let $f_*:QCS_X \to QCS_Y$ and $f^*:QCS_Y \to QCS_X$ be the usual functors. Suppose $f^*f_*:QCS_X \to QCS_X$ is naturally equivalent to the identity functor. Then f is an isomorphism.

Proof: Let $\pi:U \to X$ be an etale covering of X by an affine scheme U. X is separated so π is an affine map, so $U = \underset{\sim}{\mathrm{Spec}}\, \pi_*\mathcal{O}_U$. Then applying successively the fact that $f^*f_* = 1$, Proposition 5.2, and the fact that Y is affine, we have $U = \underset{\sim}{\mathrm{Spec}}\, \pi_*\mathcal{O}_U = \underset{\sim}{\mathrm{Spec}}\, f^*f_*\pi_*\mathcal{O}_U$ $= X \underset{Y}{\times} \underset{\sim}{\mathrm{Spec}}\, f_*\pi_*\mathcal{O}_U = X \underset{Y}{\times} \underset{\sim}{\mathrm{Spec}}\, \Gamma(Y, f_*\pi_*\mathcal{O}_U) = X \underset{Y}{\times} \underset{\sim}{\mathrm{Spec}}\, \Gamma(U,\mathcal{O}_U) = X \underset{Y}{\times} U$.

Thus we have a cartesian diagram:

$$\begin{array}{ccc} U & \xrightarrow{\;1_U\;} & U \\ \downarrow & & \downarrow \\ X & \xrightarrow[\;f\;]{} & Y \end{array}$$

We fill out one more cartesian square and label the maps

$$\begin{array}{ccccc}
U \underset{X}{\times} U & \xrightarrow{\pi_2} & U & \xrightarrow{1} & U \\
\downarrow{\scriptstyle \pi_1} & & \downarrow{\scriptstyle \pi} & & \downarrow{\scriptstyle f\pi} \\
U & \xrightarrow{\pi} & X & \xrightarrow{f} & Y
\end{array}$$

The outer corners give a cartesian square of affine schemes with base $f\pi$ faithfully flat, and π_1 etale surjective. By flat descent theory for <u>affine</u> schemes, $f\pi$ is etale surjective. Finally, since the pullback of f by an etale surjective map $f\pi$ is an isomorphism, so is f. ∎

<u>Corollary 5.4</u>: Let $f:X \to Y$ be a faithfully flat map of algebraic spaces. Then f is a categorical epimorphism.

<u>Proof</u>. The assertion is local on Y so we can take Y to be an affine scheme. Let Z be an algebraic space and $Y \underset{b}{\overset{a}{\rightrightarrows}} Z$ two maps such that $af = bf$. We must show $a = b$.

Let $Y \xrightarrow{a \times b} Z \times Z$ be the product map and $Z \xrightarrow{\Delta} Z \times Z$ the the usual diagonal map. Let $W = Y \underset{(Z \times Z)}{\times} Z$:

$$\begin{array}{ccccc}
W & \longrightarrow & Z & & \\
\downarrow{\scriptstyle \Delta'} & & \downarrow{\scriptstyle \Delta} & & \\
Y & \xrightarrow{a \times b} & Z \times Z & \underset{\pi_2}{\overset{\pi_1}{\rightrightarrows}} & Z
\end{array}$$

Then $a = b$ if and only if Δ' is an isomorphism. Note in general

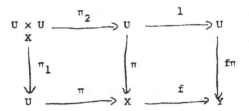

Δ is quasiaffine so Δ' is in particular quasicompact and separated. Now put in the map $f:X \to Y$ and let $W' = X \times_Y W$

$$\begin{array}{ccccccc}
W' & \longrightarrow & W & \longrightarrow & Z & & \\
\downarrow{\scriptstyle \Delta''} & & \downarrow{\scriptstyle \Delta'} & & \downarrow{\scriptstyle \Delta} & & \\
X & \xrightarrow{f} & Y & \xrightarrow{a \times b} & Z \times Z & \rightrightarrows & Z
\end{array}$$

Δ'' is an isomorphism since $af = bf$. Thus Δ'' is faithfully flat and so since f is faithfully flat, this implies that Δ' is faithfully flat. Using the fact that Δ'' is an isomorphism, a simple diagram chase shows that $\Delta'^{*}\ \Delta'_{*}:QCS_W \to QCS_W$ is naturally equivalent to the identity functor. By the sheaf criterion for isomorphism, Δ' is an isomorphism, which implies that $a = b$. ∎

<u>Construction 5.5</u>: A map $f:Y \to X$ of algebraic spaces is a <u>Stein map</u> if f is quasicompact and separated and the natural map $\mathcal{O}_X \to f_*\mathcal{O}_Y$ is an isomorphism.

Let $f:Y \to X$ be any quasicompact separated map of algebraic spaces. Then $f_*\mathcal{O}_Y$ is a quasicoherent $\mathcal{O}_X$-algebra. Let $X' = \underset{\sim}{\mathrm{Spec}}\ f_*\mathcal{O}_Y$. Then $f:Y \to X$ factors through the natural affine map $d:X' \to X$

$$\begin{array}{ccc}
Y & \xrightarrow{f'} & X' \\
{\scriptstyle f}\searrow & & \swarrow{\scriptstyle d} \\
& X &
\end{array}$$

f' is a Stein map. Given any factorization of $f:Y \xrightarrow{g} Z \xrightarrow{h} X$, with h an affine map, there is a unique map $q:X' \to Z$ with $qf' = g$ and $hq = d$.

This factorization of a quasicompact separated map f into a Stein map followed by an affine map is called the <u>Stein Factorization</u> of f. We call X' the <u>affine image</u> of f.

<u>Proposition 5.6</u>: Let $f:Y \to X$ be a Stein map of algebraic spaces. Then f is a categorical epimorphism in the category of separated algebraic spaces.

<u>Proof</u>. Let $X' \to X$ be an etale covering of X. To show that f is an epimorphism, it would be sufficient to show that the induced map $Y \times X' \to X'$ is an epimorphism. Thus we can assume that X is affine. Let a and b be two maps $X \to T$, with T a separated algebraic space, and suppose $af = bf$. We must show $a = b$.

Since T is separated and X is affine, a and b are affine maps. Hence it is sufficient to find an $\mathcal{O}_T$-algebra isomorphism $a_* \mathcal{O}_X \cong b_* \mathcal{O}_X$. But $a_* \mathcal{O}_X = a_* f_* \mathcal{O}_Y \cong b_* f_* \mathcal{O}_Y = b_* \mathcal{O}_X$. ∎

<u>Definition 5.7</u>: Let $f:Y \to X$ be a quasicompact immersion of algebraic spaces. Then f is separated so we can form the quasicoherent sheaf $f_* \mathcal{O}_Y$. Let I be the ideal making $0 \to I \to \mathcal{O}_X \to f_* \mathcal{O}_Y$ exact and let $\bar{Y} = \underset{\sim}{\text{Spec}}\, \mathcal{O}_{X/I}$. We say $\bar{Y}$ is the <u>geometric closure</u> of the subspace Y of X. f is <u>geometrically dense</u> if $\bar{Y} = X$

If $f:Y \to X$ is any immersion, we say f is <u>topologically dense</u> if for every nonempty open subspace U of X, $U \times_X Y$ is nonempty.

<u>Proposition 5.8</u>: Let $f:X \to Y$ be an immersion of separated algebraic spaces.

a) If $f:X \to Y$ is geometrically dense, f is a categorical epimorphism in the category of separated algebraic spaces.

b) If f is geometrically dense, f must be topologically dense.

<u>Proof</u>. b) is clear. To see a), note that just as in the proof of 5.2b, it is sufficient to show that if $W \to Y$ is a closed subscheme with $W \times_Y X \to X$ an isomorphism, then $W \to Y$ is an isomorphism. This assertion is local on Y and for Y affine, X and W are schemes and for schemes, this statement is EGA I.9.5.6. ∎

<u>Proposition 5.9</u>: Let X be a quasicompact locally separated algebraic space. Then there is an affine scheme U and an open immersion $i:U \to X$ such that

a) i is topologically dense

b) the induced map $U \to U \times X$ is a closed immersion

c) i is an affine map

<u>Proof</u>. Let $\pi: Y \to X$ be an etale cover of X by an affine scheme Y. By I.5.20 there is a topologically dense affine open subspace U of X such that the map $V = U \underset{X}{\times} Y \to U$ is finite as well as etale. To see b), consider the cartesian diagrams

$$\begin{array}{ccc} V & \longrightarrow & U \times Y \\ \downarrow & & \downarrow \\ U & \longrightarrow & U \times X \\ \downarrow & & \downarrow \\ X & \longrightarrow & X \times X \end{array}$$

Since X is locally separated, $U \to U \times X$ and $V \to U \times Y$ are immersions. To show $U \to U \times X$ is a closed immersion, it is sufficient to show that $V \to U \times Y$ is a proper map of schemes. But this map factors $V \to V \times Y \to U \times Y$ where the first map is a closed immersion, so proper, and the second map is finite, so proper.

c) is a simple consequence of b) since i is the composite $U \to U \times X \to X$. ∎

<u>Construction 5.10</u>: Let X be an algebraic space. Let N_X be the sheaf of nilpotents of the structure sheaf $\mathcal{O}_X$. We define X_{red}, the associated reduced space of X as $\underset{\sim}{\mathrm{Spec}}\ \mathcal{O}_{X/N_X}$.

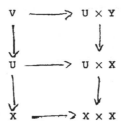

The natural map $X_{red} \to X$ is universal for maps of reduced algebraic spaces to X, so the transformation $X \rightsquigarrow X_{red}$ is a left exact functor.

Proposition 5.11: Let $f:Y \to X$ be an immersion of algebraic spaces. Then f is topologically dense if and only if $f_{red}:Y_{red} \to X_{red}$ is topologically dense. ∎

Construction 5.12: Complements of subspaces. Let X be an algebraic space and $U \to X$ an open immersion. Then the construction of the reduced closed complement of U is a local construction. Also, if $C \to X$ is a closed subspace, the construction of the open complement of C is local.

Construction 5.13: The lattice of subspaces. Given any finite collection of reduced subspaces of an algebraic space X, we can form the union or intersection of the members of the collection, since this is a local construction. Similarly, infinite unions of open subspaces and infinite intersections of reduced closed subspaces exist.

Hence we can define X to be irreducible if the intersection of any two nonempty open subspaces of X is nonempty.

Construction 5.14: Supports. Let X be a noetherian algebraic space and F a coherent sheaf on X. For each covering map $f:Y \to X$, let $\phi(Y)$ be the reduced closed subscheme of Y which is the support of f^*F. Then ϕ is a local construction, and $\phi(X)$ is called the support of F, denoted Supp(F).

<u>Proposition 5.15</u>: Let F be a coherent sheaf on a noetherian algebraic space X and J a sheaf of ideals. Suppose the closed subspace Supp(F) is contained in the closed subspace $\underset{\sim}{\mathrm{Spec}}\ \mathcal{O}_{X/J}$. Then there is an integer n such that $J^n F = 0$.

<u>Proof</u>. Let $f:Y \to X$ be a covering of X, with Y affine. Then it is sufficient to show that $(f^*J)^n (f^*F) = 0$ for some n and this is EGA I.9.3.4. ∎

<u>Corollary 5.16</u>: In the situation above, let $C = \underset{\sim}{\mathrm{Spec}}\ \mathcal{O}_{X/J^n}$ and $i:C \to X$ the canonical injection. Then $F = i_* i^* F$. ∎

<u>Definition 5.17</u>: A property P of algebraic spaces is <u>inductive</u> if P is true of the empty space, and for any algebraic space X, if P(Y) holds for every closed subspace Y of X with $Y_{red} \neq X_{red}$, then P(X) holds.

<u>Proposition 5.18</u>: (Noetherian Induction) Let X be a noetherian algebraic space and P an inductive property. Then P(X) holds.

<u>Proof</u>. Let $U \to X$ be an etale covering of X with U an affine noetherian scheme. Then X has no infinite descending chain of closed subspaces, since such would lift to an infinite descending chain of closed subspaces of U, so an infinite ascending chain of ideals in the noetherian ring $\Gamma(U, \mathcal{O}_U)$. Hence if P(X) fails, there must be a minimal closed subspace $Y \subseteq X$ where P(Y) fails. We can assume Y = X. Then for

every closed subspace X' of X, $X'_{red} \neq X_{red}$, $P(X')$ holds. Hence $P(X)$. ∎

6. Points and the Zariski Topology

Definition 6.1: Let X be an algebraic space. A point of X is a map of algebraic spaces, $i:p \to X$, where p is the spectrum of a field k and i is a categorical monomorphism. By abuse of language, we say "p is in X" and write $p \in X$. k is the residue field of X at p and will sometimes be written $k(p)$. Two points $i_1:p_1 \to X$ and $i_2:p_2 \to X$ are equivalent if there is an isomorphism $e:p_1 \to p_2$ with $i_2 e = i_1$. Again by abuse of language we consider two equivalent points to be identical.

A geometric point $j:q \to X$ is any map of algebraic spaces with q the spectrum of a separably closed field. (Note that a geometric point is usually not a point.)

Proposition 6.2: Let $f:q \to X$ be a map of algebraic spaces where $q = \text{Spec } k$ for some field k. Then there is a point p of X such that f factors $q \to p \to X$.

Proof. By 3.13, X has a covering $\{X_i \xrightarrow{\varphi_i} X\}$ by quasicompact open subspaces. $f:q \to X$ must factor $q \to X_i \to X$ for some i. Hence we can assume that X is quasicompact.

For any covering map $Y \to X$, with Y a quasicompact scheme, let $\emptyset(Y)$ be the (finite!) underlying set of points of $q \times_X Y \to Y$. This is a local construction. Let $p = \emptyset(X)$. Since p is the quotient of an etale equivalence relation $\emptyset(Y \times_X Y) \rightrightarrows \emptyset(Y)$, where each of $\emptyset(Y \times_X Y)$ and $\emptyset(Y)$ are finite disjoint unions of affine spectra of fields, p is a disjoint union of affine spectra of fields. (See I.5.15). $q \times_X Y \to \emptyset(Y)$ is surjective so $q \to p$ is surjective. Hence p is the affine spectrum of a field. Finally, $\emptyset(Y) = p \times_X Y \to Y$ is a monomorphism so $p \to X$ is a monomorphism. ∎

Corollary 6.3: Every algebraic space $X \neq \emptyset$ has at least one point.

Proof. Let $Y \to X$ be a covering map of X by a scheme Y and $q \to Y$ a point of Y. By 6.2, there is a point $p \to X$ and a map $q \to p$ such that $q \to p \to X = q \to Y \to X$. ∎

Theorem 6.4: Let X be an algebraic space and $x \longrightarrow X$ a point of X. Then there is an affine scheme U and an etale map $U \longrightarrow X$ such that $x \longrightarrow X$ factors $x \longrightarrow U \longrightarrow X$.

(The proof uses the notion of symmetric product and will be deferred until IV.2.6, p.190. We have no need for this fact but it is sometimes useful - for example in the construction of quotients of algebraic spaces by finite group actions (IV.1.8).) ∎

Definition 6.5: A point $p \to X$ is scheme-like if there is an affine scheme U and an open immersion $U \to X$ such that $p \to X$ factors $p \to U \to X$.

Proposition 6.6: Let X be an algebraic space. Then there is an open subspace U of X such that U is a scheme, and a point $p \to X$ is in U if and only if p is scheme-like. X is a scheme if and only if all its points are scheme-like. ∎

We sometimes say that, "U is the open subspace where X is a scheme."

Proposition 6.7: Let X be an algebraic space. Then X is a scheme almost everywhere, i.e., the open subspace of scheme-like points is topologically dense.

Proof. By 3.13, X has a covering $\{X_i \xrightarrow{\varphi_i} X\}$ with each φ_i an open immersion. The scheme-like points in X are dense iff they are dense in each X_i. Hence we can assume X is quasicompact. Let $R \rightrightarrows U$ be an etale equivalence relation of quasicompact schemes with quotient X. By I.5.20, there is a dense open subspace V of U, such that the induced etale equivalence relation $S = R \underset{(U \times U)}{\times} (V \times V) \rightrightarrows V$ is effective in the category of schemes. Let Y be the quotient. Then $Y \to X$ is a dense open subscheme of X. The open subspace U of X where X is a scheme contains Y. ∎

<u>Corollary 6.8</u>: Let X be a noetherian algebraic space. Then there is a topologically dense open immersion $U \to X$ with U an affine scheme.

<u>Proof:</u> Using Proposition 6.7 we can replace X by its dense open subscheme of scheme-like points. I.e., we can assume X is a scheme, and then such a U is easily constructed. (Specifically, pick $U_1 \subset X$ open and affine. If U_1 is not dense, pick $U_2 \subset X-U_1$, open and affine. If $U_1 \cup U_2$ is not dense, pick $U_3 \subset X-U_1-U_2$, etc. X is noetherian, so this procedure must terminate after, say, n steps. $U = U_1 \cup \cdots \cup U_n$ is a finite disjoint union of affine schemes, hence affine, and by construction U is topologically dense.) ∎

<u>Definition 6.9:</u> Let X be an algebraic space. The <u>associated underlying topological space</u> of X, $|X|$, is defined as the collection of points of X (modulo equivalence of points). The set $|X|$ is given a topological structure by taking a subset $Q \subset |X|$ to be closed if Q is of the form $|Y|$ for some closed subspace Y of X. By abuse of language the topology on $|X|$ is called the <u>Zariski topology</u> on X.

<u>Proposition 6.10:</u> $|X|$ is a topological space and there is a one-one correspondence between open subspaces of X and open subsets of $|X|$, and a one-one correspondence between reduced closed subspaces of X and closed subsets of $|X|$. Also $X \rightsquigarrow |X|$ is a functor.

Proof: By 5.13, X is a topological space. Let U_1, U_2 be two different subspaces of X; say the complement C of U_1 in U_2 is nonempty. C is a nonempty subspace so has a point q by 6.3 and $q \in U_2 - U_1$ so $|U_2| \neq |U_1|$. Similarly for reduced closed subspaces.

If $f:X \to Y$ is a map of algebraic spaces and $q \to X$ a point, then by 6.2 there is a point $p \to Y$ and a map $q \to p$ such that $q \to X \to Y = q \to p \to Y$. ∎

6.11: Note that X is irreducible iff $|X|$ is and our previous definitions of topologically dense subspace, surjective map, universally open map, open map and immersion are equivalent to the usual definitions via $|X|$. We say that a map of algebraic spaces $Y \to X$ is closed if the associated map $|Y| \to |X|$ is closed, and universally closed if for every algebraic space $X' \to X$ mapping to X, the induced map $|Y \times_X X'| \to |X'|$ is closed.

6.12: $|X|$ can in fact be considered as a local ringed space, taking for an open subset $|U|$ of $|X|$, $\Gamma(|U|, \mathcal{O}_{|X|}) = \Gamma(U, \mathcal{O}_X)$. The reader should note carefully that this functor, restricted to locally separated algebraic spaces, is neither faithful nor full. (See Example 2 in the Introduction.)

Construction 6.13: Atomizations. An atom is an affine scheme SpecR where R is a complete local ring with separably closed residue field.

Let X be an algebraic space. For each point p of X, with residue field k(p), let q be the affine spectrum of the separable closure of k(p) and $i: q \to X$ the natural map. Let $\bar{\bar{X}}_p = \operatorname{Spec} \widehat{\Gamma(q, i^* \mathcal{O}_X)}$. Then $\bar{\bar{X}}_p$ is an atom, the <u>atom of X at p</u>. Let $\bar{\bar{X}}$ be the disjoint union, $\bar{\bar{X}} = \bar{\bar{X}}_p$, over all points p of X. $\bar{\bar{X}}$, or more specifically, the map $\bar{\bar{X}} \to X$ is called the <u>atomization</u> of X. X is <u>atomic</u> if $X = \bar{\bar{X}}$.

<u>Proposition 6.14:</u> Let X be an algebraic space.

a) For any point p in X, the map $\bar{\bar{X}}_p \to X$ is flat.

b) The map $\bar{\bar{X}} \to X$ is faithfully flat.

c) The map $\bar{\bar{X}}$ is a categorical epimorphism in the category of algebraic spaces.

<u>Proof:</u>

a) Let $\pi: Y \to X$ be an etale covering with Y a disjoint union of affine schemes. Let q be a point of Y with $\pi(q) = p$. The map $\bar{\bar{X}}_p \to X$ factors $\bar{\bar{X}}_p \to Y$ identifying $\bar{\bar{X}}_p$ with $\bar{\bar{Y}}_q$. Hence we can assume X is an affine scheme where this is well known.

b) Again, one reduces easily to the case of affine X where this is well known.

c) is a corollary of b) and 5.4. ∎

We now apply the technique of atomization to prove the following theorem. This is a weak version for algebraic spaces of the stronger Deligne Theorem for schemes, mentioned previously. Note the result is not a trivial consequence of its validity for schemes.

Theorem 6.15: Let $f:X \to Y$ be a map of algebraic spaces which is quasifinite, of finite presentation and separated. Then f is quasiaffine.

Proof: The assertion is local on Y so we can assume Y is affine: Y = Spec R. R can be considered as a direct limit of its noetherian subrings R_i and be taking a sufficiently large R_i, a noetherian scheme Y_o = Spec R_i can be found with a map $Y \to Y_o$, and a quasifinite separated map locally of finite presentation $X_o \to Y_o$ so that $X = X_o \times_{Y_o} Y$. Once we show $X_o \to Y_o$ is quasiaffine, it follows that $X \to Y$ is quasiaffine.

Also since f is quasicompact and separated, we can take its Stein factorization $X \xrightarrow{f_1} X_1 \xrightarrow{d} Y$ where $f_{1*}\mathcal{O}_X = \mathcal{O}_{X_1}$ and d is affine. Applying I.1.21, f_1 is quasifinite, of finite presentationand separated. We are thus reduced to showing the following lemma:

Lemma 6.15a: Let $f:X \to Y$ be a quasifinite separated Stein map of finite presentationwith Y a noetherian affine scheme. Then f is an open immersion.

Proof: We use the technique of atomization to show that f is flat and a monomorphism. First we deal with the case where Y is an atom and the image of the map f contains the closed point of Y.

Let $X' \to X$ be a (quasifinite) etale covering of X by an affine scheme X'. By Zariski's Main Theorem (for affine schemes--see EGA III.4.4.3) X' is isomorphic to an open subspace of an affine scheme

Y' finite over Y. Since Y is an atom, Y' is a finite disjoint sum of atoms (Hensel's Lemma) so X' is a finite disjoint sum of subspaces. Since at least one point of X' maps onto the closed point of Y, one of the components X_1 of X' must be isomorphic to a component of Y'. The map $X_1 \to Y$ is finite so $X_1 \to X$ is finite. Hence $X_1 \to X$ is proper so its image is closed. But $X_1 \to X' \to X$ is etale so its image is open. Thus X_1 maps onto a component of X. Since $X \to Y$ is Stein, and Y is connected (being an atom) this component must be all of X. Since $X_1 \to X$ is etale and finite and X_1 is affine, so is X (by 1.5.14). Hence $X \to Y$ is a Stein map of affine schemes, hence an isomorphism.

In the case of general $f: X \to Y$, let p be a point in Y such that $f^{-1}(p)$ is nonempty. Consider the cartesian diagram:

$$\begin{array}{ccc} X \times_Y \overline{\overline{Y}}_p & \longrightarrow & \overline{\overline{Y}}_p \\ \downarrow & & \downarrow \\ X & \xrightarrow{f} & Y \end{array}$$

Since $\overline{\overline{Y}}_p \to Y$ is flat (6.14b), the top map is a stein map. (We are here also applying 4.18). $X \times_Y \overline{\overline{Y}}_p \to \overline{\overline{Y}}_p$ is also quasifinite and separated, so the map must be an isomorphism by the above argument in the special case. Now let Y_1 be the disjoint union of all $\overline{\overline{Y}}_p$ for all points p in Y for which $f^{-1}(p)$ is nonempty. We have a cartesian diagram

$$\begin{array}{ccc} \overline{\overline{X}} & \xrightarrow{f'} & Y_1 \\ \varphi \downarrow & & \downarrow \alpha \\ X & \xrightarrow{f} & Y \end{array} \qquad \text{with } f' \text{ an isomorphism}$$

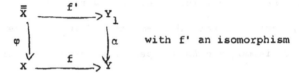

By 6.14, φ is faithfully flat, and α is flat. f' is an isomorphism so faithfully flat. Hence $f:X \to Y$ is flat.

Also f is a monomorphism. To see this, let $Z \overset{a}{\underset{b}{\rightrightarrows}} X$ be two maps with $fa = fb$. Let $Z_1 = Z \underset{Y}{\times} Y_1$. We have a diagram

$$\begin{array}{ccccc} Z_1 & \overset{a'}{\underset{b'}{\rightrightarrows}} & \bar{\bar{X}} & \overset{f'}{\longrightarrow} & Y_1 \\ \downarrow{\scriptstyle \varphi'} & & \downarrow{\scriptstyle \varphi} & & \downarrow{\scriptstyle \alpha} \\ Z & \overset{a}{\underset{b}{\rightrightarrows}} & X & \overset{f}{\longrightarrow} & Y \end{array}$$

where $\varphi a' = a\varphi'$, $\varphi b' = b\varphi'$. Since f' is an isomorphism, it is monic. Hence $a'f' = b'f'$ implies $a' = b'$ implies $\varphi a' = \varphi b'$ implies $a\varphi' = b\varphi'$. By 6.14, φ is faithfully flat so φ' is faithfully flat, so by 5.4, φ' is an epimorphism. Hence $a = b$.

To finish our theorem we only need now show the following lemma:

<u>Lemma 6.15b:</u> Let $f:X \to Y$ be a flat monomorphism of finite presentation. Then f is an open immersion.

<u>Proof:</u> Since f is flat and of finite presentation, it is open. (By 3.5). Hence the image of f is an open subspace of Y so without loss of generality, we can assume f is surjective, hence faithfully flat. Since f is monic, $X \underset{Y}{\times} X = X$ so f is nonramified. Thus f is an etale surjective monomorphism.

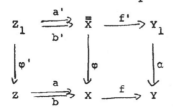

X is the quotient of the etale equivalence relation $X \overset{1}{\underset{1}{\rightrightarrows}} X$ and Y is the quotient of the etale equivalence relation. $X \underset{Y}{\times} X \rightrightarrows X$. Since $X \underset{Y}{\times} X = X$, the induced map $X \to Y$ (which is our original f) must be the identity map. ∎

<u>Corollary 6.16:</u> Let $f:X \to Y$ be a map of algebraic spaces which is locally quasifinite, locally of finite presentation and separated. Suppose Y is a scheme. Then X is a scheme.

<u>Proof:</u> By 3.13, there is a covering $\{X_i \overset{\varphi_i}{\to} X\}$ of X by quasicompact algebraic spaces X_i, with each φ_i an open immersion and such that X is a scheme iff each X_i is a scheme. Hence we can assume X is quasicompact so f is quasicompact. Applying the theorem 6.15, f is quasiaffine, so by 3.8, X is a scheme. ∎

<u>Corollary 6.17:</u> Let $f:X \to Y$ be an etale separated map of algebraic spaces. If Y is a scheme, then so also is X. ∎

7. Proper and Projective Morphisms

Definition 7.1: A map $f:X \to Y$ of algebraic spaces is proper if f is separated, of finite type, and universally closed. f is finite if f is affine and $f_* \mathcal{O}_X$ is a coherent $\mathcal{O}_Y$-module.

Proposition 7.2: The classes of proper morphisms and finite morphisms are stable in the etale topology. A closed immersion is finite. A finite morphism $f:X \to Y$ is proper.

Proof: The first two assertions are straightforward. The last is local on Y so we can assume that Y, and hence X, is affine. This is then EGA II.6.1.10. ∎

7.3: A fundamental theorem--in fact the main theorem of this book--is the Finiteness Theorem: If $f:X \to Y$ is a proper morphism of noetherian algebraic spaces, and F is a coherent sheaf on X, then all the higher direct images $R^q f_* F$ are coherent $\mathcal{O}_Y$-modules.

The major use of this theorem is when X is a "variety over a separably closed field k". Here Y = Spec k, $f:X \to Y$ is proper, and X is, say, reduced and irreducible. Then one defines the usual numerical invariants of X by taking the dimensions of the vector spaces associated with the coherent sheaves $R^q f_* F$ for various canonical coherent sheaves F on X.

In one case the finiteness theorem is obvious. If f is a finite morphism, $f_* \mathcal{O}_X$ is a coherent $\mathcal{O}_Y$-module so f_* preserves coherent sheaves. Since f is affine, all the $R^q f_* F = 0$ for $q > 0$, so these are certainly coherent. The general case of the theorem involves considerably much more work and we will only be able to prove it after setting up two chapters of machinery.

Proposition 7.4: Let $f:X \to Y$ be a finite etale morphism with X and Y noetherian. Then $f_* \mathcal{O}_X$ is a locally free $\mathcal{O}_Y$-module. Suppose Y is irreducible. Then the rank of $f_* \mathcal{O}_X$ is constant (and is called the degree of f).

Proof: For the local freeness, it is sufficient to assume Y is the affine spectrum of a local ring. Since f is finite and etale, $f_* \mathcal{O}_X$ is a finite and flat $\mathcal{O}_Y$-module, hence free.

For the second assertion, note the Nakayama lemma implies that $f_* \mathcal{O}_X$ has constant rank in an open neighborhood of any point. ∎

7.5: We now define projective and quasi-projective morphisms. There are several ways these can be defined and the detailed theory involves such notions as the Proj construction and ample sheaves. (See EGA II) Our definition however, will entail that, if $f:X \to Y$ is a projective or quasi-projective morphism, and Y is a scheme, then X is a scheme. Thus any projective construction in algebraic spaces over a base scheme is the same as the scheme-theory case. We will give an indication of stability of projective morphisms in the etale topology but otherwise leave to the reader the task of transcribing the detailed theory of projective morphisms.

<u>Definition 7.6:</u> Let $\mathbb{P}^n_{\mathrm{Spec}\ Z}$ be projective n-space over the affine spectrum of the ring of integers Z. (We assume the reader is familiar with this object.) Let X be any algebraic space. We define $\mathbb{P}^n_X$ as $\mathbb{P}^n_{\mathrm{Spec}\ Z} \underset{(\mathrm{Spec}\ Z)}{\times} X$, <u>projective n-space over X</u>. There is a canonical invertible sheaf $\mathcal{O}(1)$ on $\mathbb{P}^n_{\mathrm{Spec}\ Z}$ and we use the map $\mathbb{P}^n_X \to \mathbb{P}^n_{\mathrm{Spec}\ Z}$ to induce a canonical invertible sheaf on $\mathbb{P}^n_X$, which we also denote $\mathcal{O}(1)$.

A map $f:Y \to X$ of algebraic spaces is <u>quasiprojective</u> if there is an integer n and an immersion $i:Y \to \mathbb{P}^n_X$ such that f is the composite $Y \xrightarrow{i} \mathbb{P}^n_X \to X$. f is <u>projective</u> if for some such i, i is a closed immersion. Note (applying 3.8) if $f:Y \to X$ is quasiprojective and X is a scheme, then so also is Y.

<u>Proposition 7.7:</u> The classes of projective and quasiprojective morphism of algebraic spaces are closed subcategories. ∎

(Note they are not stable in the etale topology. Indeed the class of projective maps of schemes is not stable in the category of schemes in the Zariski topology.)

<u>Proposition 7.8:</u> A quasiprojective map is separated and of finite type. A quasiprojective map is projective if and only if it is proper.

<u>Proof:</u> The only hard part of this is proving that the map $\mathbb{P}^n_X \to X$ is proper which comes down to showing that the map $\mathbb{P}^n_{\mathrm{Spec}\ Z} \to \mathrm{Spec}\ Z$ is proper. This is proved in EGA II.5.5.3. ∎

<u>Definition 7.9:</u> Let $f:Y \to X$ be a map of algebraic spaces. Let $\mathcal{L}$ be an invertible sheaf on Y. We say $\mathcal{L}$ is <u>f-ample</u> if f is quasi-projective, and for some factorization of f, $Y \xrightarrow{i} \mathbb{P}^n_X \to X$, with i an immersion, and for some integer k, the k-fold tensor product, $\mathcal{L}^{\otimes k}$, of $\mathcal{L}$ is isomorphic to $i^*\mathcal{O}(1)$ where $\mathcal{O}(1)$ is the canonical sheaf on $\mathbb{P}^n_X$.

As in scheme theory, if $f:Y \to X$ is a map of algebraic spaces, and an invertible $\mathcal{O}_Y$-module $\mathcal{L}$ is f-ample, one can reconstruct from $\mathcal{L}$ an appropriate imbedding $Y \xrightarrow{i} \mathbb{P}^n_X$ so $\mathcal{L} = i^*\mathcal{O}(1)$. (Since, given $\mathcal{L}$, the construction is local on X so we can take X, hence Y to be a scheme.)

<u>Proposition 7.10:</u> Let

$$\begin{array}{ccc} U & \xrightarrow{h} & Y \\ {\scriptstyle g}\downarrow & & \downarrow{\scriptstyle f} \\ V & \xrightarrow{\pi} & X \end{array}$$

be a cartesian diagram of algebraic spaces with π etale and surjective. Suppose g is quasi-projective and suppose further that there is an invertible sheaf $\mathcal{L}$ on Y such that $h^*\mathcal{L}$ is g-ample. Then $\mathcal{L}$ is f-ample, so f is quasi-projective.

<u>Proof:</u> See EGA IV.2.7.2 ∎

<u>Theorem 7.11:</u> (The Serre Finiteness Theorem). [FAC XXIX] Let $f:X \to Y$ be a projective morphism with $\mathcal{L}$ an ample sheaf on X. Let F be a coherent sheaf on X. We write $F(n)$ for $F \otimes \mathcal{L}^{\otimes n}$.

Then

(1) $R^q f_*(F)$ is coherent, for all $q \geq 0$.

(2) There is an integer N, such that for all $n \geq N$ and $q > 0$, $R^q f_*(F(n)) = 0$.

(3) There is an integer N, such that for all $n \geq N$, the natural map $f^* f_* F(n) \to F(n)$ is surjective.

Proof: The assertions are all local on Y in the following sense. Let $E \xrightarrow{\pi} Y$ be an affine etale cover of Y. Then there are cartesian diagrams

$$\begin{array}{ccc} E \times_Y X & \xrightarrow{\pi_2} & X \\ \downarrow & & \downarrow \\ \mathbb{P}^n_E & \xrightarrow{\pi_1} & \mathbb{P}^n_Y \\ \downarrow & & \downarrow \\ E & \xrightarrow{\pi} & Y \end{array}$$

with $\pi_2^*(F(m)) = \pi_2^*(F)(m)$ and, letting $f' : E \times_Y X \to E$ be the pullback of $f : X \to Y$, $R^q f'(\pi_2^*(F(m))) = \pi^*(R^q f(F(n)))$ (see II.4.17). Thus we are reduced to the case where Y is affine. Then by the remark in 7.6, X is a scheme and the assertion is precisely EGA III, 2.2.1. ∎

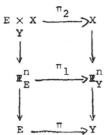

8. Integral Algebraic Spaces

Definition 8.1: Let X be a quasicompact locally separated algebraic space. We say X is integral if X is also irreducible and reduced.

8.2: Let X be an integral algebraic space. If $i:U \to X$ is an open immersion with U quasicompact, then U is also integral. If U is affine, then U is the spectrum of an integral domain. On the other hand, if $Y \to X$ is an etale map, Y need not be integral. (Y must be reduced of course.)

Proposition-Definition 8.3: Let X be an integral algebraic space. Then there is a unique point $x_o \in X$ such that the closure of x_o is X. (I.e., the only reduced closed subspace of X containing x_o is X.) x_o is called the generic point of X. x_o is a scheme-like point of X. The map $x_o \to X$ is affine.

Proof: By 5.9, there is a dense affine open subspace U of X. Clearly any such point must lie in U. As mentioned above, U is the affine spectrum of an integral domain. Then the point of U corresponding to the zero ideal of $\Gamma(U, \mathcal{O}_X)$ is the desired x_o. By 5.9 U can be chosen so that the map $U \to X$ is affine. Then $x_o \to U \to X$ is an affine map. ∎

Definition 8.4: A map $f:Y \to X$ of algebraic spaces is birational if there is a (topologically) dense open subspace U of X such that the restriction of f to $Y \times_X U \to U$ is an isomorphism.

Construction 8.5: (Decomposition of a noetherian locally separated algebraic space into its components.) Let X be a noetherian locally separated algebraic space. The associated topological space $|X|$ of X is then a noetherian topological space. I.e., there are no infinite descending chains of closed subsets of $|X|$.

We claim that $|X|$ is the union of a finite number of maximal irreducible closed subsets. To see this, first note that any point of $|X|$ is contained in an irreducible closed subset (the closure of the point) and every irreducible closed subset is contained in a maximal irreducible closed subset (apply the Zorn lemma for this). Thus $|X|$ is the union of the set of distinct maximal irreducible closed subsets $\{C_i\}_{i \in I}$ of X. For each such C_i, let $V_i \to X$ be the reduced algebraic subspace of X with $|V_i| = C_i$. Each V_i is integral so admits a generic point $x_i \to V_i$. C_i is the closure of its generic point x_i. Also $x_i \in C_j$ iff $i = j$. (Since otherwise $C_i \subseteq C_j$ contradicting the maximality and distinctness of the C_i's). Hence if the set $\{C_i\}_{i \in I}$ is infinite we can pick a countable subset $I' = \{1,2,3,\ldots\} \subseteq I$ and obtain an infinite descending chain of closed subspaces of $|X|$: $|X| \geq \overline{|X| - C_1} \geq \overline{|X| - (C_1 \; C_2)} \geq \cdots$ (where $\overline{}$ denotes closure). Each inclusion here is proper since

$x_i \in \overline{|X| - C_1 \ldots C_{i-1}}$) but not in $\overline{|X| - (C_1 \ldots C_i)}$. This contradicts the noetherian hypothesis. Hence the index set I is finite. Say $I = \{1,2,\ldots,n\}$.

Let $U_i = |X| - (C_1 \cup \ldots \cup C_{i-1} \cup C_{i+1} \ldots \cup C_n)$. U_i is an open subspace of $|X|$ since the C_i are closed. Let now for each i, $\varphi_i : X_i \to X$ be an open subspace with $|X_i| = U_i$. φ_i is quasicompact and separated so we can form W_i = the geometric closure of X_i in X. Let W be the disjoint union of the W_i. We then have the following proposition:

<u>Proposition 8.5:</u> Let X be a noetherian locally separated algebraic space. Then there is an algebraic space W which is a finite disjoint union of irreducible closed subspaces of X and the natural map $W \to X$ is birational, projective and surjective. If X is reduced, W is a disjoint union of integral algebraic spaces.

<u>Proof:</u> The only thing left to check is the birationality. But the open subset X which is the union of $X_i - (X_1 \cup \ldots X_{i-1} \cup X_{i+1} \cup X_n)$ satisfies the requirement.

<u>8.6:</u> We now define the notion of <u>stalks</u> of sheaves. As usual, AS_X will denote the category of abelian sheaves on an algebraic space X, and QCS_X the category of quasicoherent sheaves. If $f: X \to Y$ is a map of algebraic spaces and $Q \in AS_Y$, we write $f^{ab}Q$ for the associated inverse image of Q in AS_X. Also if $Q \in QCS_Y$, we use

$f^{ab}Q$ to denote the abelian sheaf inverse image of Q in AS_X. This convention will hold throughout this discussion, through paragraph 8.12. (See 4.4 for the distinction between the two functors $f^{ab}:AS_Y \to AS_X$ and $f^m:QCS_Y \to QCS_X$.)

The classical definition of the stalk F_x of an abelian sheaf F on X at a point $i:x \to X$ is $F_x = \Gamma(x, i^{ab}F)$. The functor (Stalk at x) $:AS_X \to Ab$ is the composite of the two functors $i^{ab}AS_X \to AS_x$ and $\Gamma(x,-):AS_x \to Ab$. In the case of topological spaces, or the case of schemes in the Zariski topology, $\Gamma(x,-)$ is exact. Since i^{ab} is always exact, this makes (Stalk at x) an exact functor. One then shows that the functor $AS_X \to Ab$, taking any abelian sheaf F to the direct sum of its stalks at all points of X is not only exact but faithful. Thus "to check that a sequence of sheaves is exact, it is sufficient to check at the stalks".

In the etale topology of schemes or algebraic spaces, the functor $\Gamma(x,-):AS_x \to Ab$ is no longer exact unless x is the spectrum of a separably closed field. This leads to the definitions of two different kinds of stalks.

Definition 8.7: Let X be an algebraic space and F an abelian sheaf on X.

(1) Let $i:x \to X$ be a geometric point of X. Then the geometric

stalk of F at x is the abelian group $\Gamma(x, i^{ab}F)$.

(2) Let $i: x \to X$ be a point of X. Then the stalk of F at x is the abelian sheaf $i^{ab}F$ on x.

(3) If F is in fact a quasicoherent sheaf on X, its stalk in either sense is its stalk as an abelian sheaf. The fiber of a quasicoherent sheaf F at a (geometric or ordinary) point $i: x \to X$ is the k(x)-module $\Gamma(x, i^{ab}F \otimes_X \mathcal{O}_x) = \Gamma(x, i^m F)$.

Proposition 8.8: Let X be an algebraic space. Then the functor (Geometric Stalk)$:AS_X \to Ab$ which takes each abelian sheaf into the sum of its stalks at all the geometric points of X is exact and faithful. ∎

Definition 8.9: Let X be an integral algebraic space with generic point $i: x_o \to X$. The function field K of X is defined by $K \neq k(x_o)$. The function field sheaf $\tilde{K}$ on X is the sheaf $i_* \mathcal{O}_{x_o}$.

8.10: With the above notation, $K = \Gamma(x_o, i^{ab}\mathcal{O}_X)$ --a fact which one can check on any open subspace U of X with $x_o \in U$, and thus check in the case where X is affine, where it is clear.

K is quasicoherent since i is quasicompact and separated. Also K is a flat $\mathcal{O}_X$-module since the inverse image functor $QCS_X \to QCS_{x_o}$ is just the restriction of $i^{ab}: AS_X \to AS_{x_o}$ which is exact. Note that unlike the corresponding definition in the theory of schemes, $\tilde{K}$ is not a constant sheaf. Indeed, if $Y \to X$ is an etale map and Y has irreducible components $Y_1, \ldots, Y_n$ with generic points

$y_1,\ldots,y_n$, then $\Gamma(Y,\tilde{K}) = \oplus k(y_i)$ - a finite separable field extension of $\Gamma(X,\tilde{K}) = K$.

Finally, note that since $\tilde{K} = i_* i^{ab} \mathcal{O}_X$, where $i:x_o \to X$ is the generic point, there is a natural inclusion $\mathcal{O}_X \to \tilde{K}$.

<u>Definition 8.11:</u> Let X be a separated integral algebraic space and $\mathcal{O}_X$ its structure sheaf and $\tilde{K}$ its sheaf of function fields. Let $\mathcal{O}_X^*$ be the sheaf of units of $\mathcal{O}_X$, and $\tilde{K}^*$ the sheaf of units of $\tilde{K}$. Then the injection $0 \to \mathcal{O}_X \to \tilde{K}$ induces an injection $0 \to \mathcal{O}_X^* \to \tilde{K}^*$ of abelian sheaves. Let $\mathcal{D}$ be the quotient (abelian) sheaf so $0 \to \mathcal{O}_X^* \to \tilde{K}^* \to \mathcal{D} \to 0$ is exact. $\mathcal{D}$ is called the sheaf of <u>Cartier divisors</u> on X. A <u>Cartier divisor</u> on X is an element of $\Gamma(X,\mathcal{D})$. The class of principal Cartier divisors is the image of $\Gamma(X,K^*)$ in $\Gamma(X,\mathcal{D})$.

<u>Theorem 8.12:</u> $H^1(X,\tilde{K}^*) = 0$. Thus there is an exact sequence of abelian groups

$$0 \to \Gamma(X,\mathcal{O}_X^*) \to \Gamma(X,\tilde{K}^*) \to \Gamma(X,\mathcal{D}) \to H^1(X,\mathcal{O}_X^*) \to 0$$

<u>Proof:</u> (In the case of schemes in the Zariski topology this is easy--$\tilde{K}^*$ is a constant sheaf. But we have something to prove.)

Let $i:x_o \to X$ be the inclusion of the generic point in X, so $\tilde{K}^* = i_*\mathcal{O}_{x_o}^*$. Then there is a spectral sequence

$$E_2^{pq} = H^p(X, R^q_{i_*}\mathcal{O}_{x_o}^*) \Rightarrow H^{p+q}(x_o, \mathcal{O}_{x_o}^*)$$

As always, this gives an injection $0 \to E_2^{10} \to H^1$ so there is a monomorphism $H^1(X,\tilde{K}^*) \to H^1(x_o, \mathcal{O}_{x_o}^*)$. (Another way to see this is to note that the Cech complexes defining $H^{\cdot}(X,\tilde{K})$ and $H^{\cdot}(x_o, \mathcal{O}_{x_o}')$ are identical.)

Hence we only need show that $H^1(x_o, \mathcal{O}_{x_o}) = 0$. But this fact is the Hilbert Theorem 90! (See SGAA IX,3.3).

Thus we have for algebraic spaces X the usual fact that $H^1(X,\mathcal{O}_X)$ classifies both locally free sheaves, and Cartier divisors modulo principal Cartier divisors. As usual, we write $H^1(X,\mathcal{O}_X^*) = \text{Pic } X$, the <u>Picard Group</u> of X.

<u>8.13:</u> In the final part of this section, we discuss coherent sheaves on a noetherian integral algebraic space X which will remain fixed throughout. We write $\mathcal{O}_X$ for its structure sheaf, $i: x_o \to X$ for the generic point and K for the function field. Recall that $\tilde{K}$, the function field sheaf, is quasicoherent.

<u>Lemma 8.13:</u> Let G be a coherent sheaf on X. Then there is a dense open subspace V of X such that the restriction of G to V is a free $\mathcal{O}_V$-module.

<u>Proof:</u> By 6.8, X has a dense affine open subspace U and we can clearly assume X = U, i.e., that X is affine. This is then a simple corollary of the Nakayama lemma. ∎

The rank of this free module is clearly independant of the choice of U and will be called the <u>generic rank</u> of G.

<u>Definition 8.14:</u> Let F be a coherent $\mathcal{O}_X$-module. Consider the map $F \to F \otimes_{\mathcal{O}_X} \tilde{K}$. Let $F_1 \to F$ be the kernel. Then F_1 is coherent and

Supp F_1 is a proper closed subspace of X. We say that F is a torsion sheaf if $F = F_1$ and that F is torsion-free if $F_1 = 0$.

Proposition 8.15:

a) A coherent sheaf F is torsion if and only if F has generic rank zero if and only if Supp $F \neq X$.

b) For any coherent sheaf F, there is an exact sequence $\mathcal{E}: 0 \to F_1 \to F \to F_2 \to 0$ with F_1 torsion and F_2 torsion-free.

c) If F is a torsion sheaf and G is a torsion-free sheaf, $\mathrm{Hom}_{QCS_X}(F,G) = 0$.

Proof:

a) is clear.

b) Given F, define F_1 as in 8.14 and let F_2 be the quotient sheaf of $F_1 \to F$. $\tilde{K}$ is a flat $\mathcal{O}_X$-module so the induced sequence $\mathcal{E} \otimes_{\mathcal{O}_X} \tilde{K}$ is exact. Consider the following diagram where we we have labeled some maps for convenience:

$$\begin{array}{ccccccccc} 0 & \longrightarrow & F_1 & \longrightarrow & F & \longrightarrow & F_2 & & 0 \\ & & \downarrow a & & \downarrow & & \downarrow b & & \\ 0 & \longrightarrow & F_1 \otimes_{\mathcal{O}_X} \tilde{K} & \longrightarrow & F \otimes_{\mathcal{O}_X} \tilde{K} & \longrightarrow & F_2 \otimes_{\mathcal{O}_X} \tilde{K} & \longrightarrow & 0 \end{array}$$

Since F_1 is torsion, $F_1 \otimes_{\mathcal{O}_X} \tilde{K} = 0$, so a is the zero map. Applying the snake lemma, Ker $b = 0$.

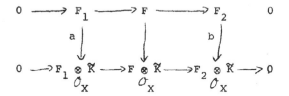

c) Let $\varphi: F \to G$ be any map. Let H be the image of φ. Supp $H \subseteq$ Supp F so H is torsion. φ is the zero map iff the inclusion $H \to G$ is the zero map. Hence we can assume φ is an injection. Consider the commutative diagram:

$$\begin{array}{ccc} F & \xrightarrow{\varphi} & G \\ {\scriptstyle a}\downarrow & & \downarrow{\scriptstyle b} \\ F \otimes_{\mathcal{O}_X} \widetilde{K} & \xrightarrow{\varphi'} & G \otimes_{\mathcal{O}_X} \widetilde{K} \end{array}$$

(where φ' is injective since $\widetilde{K}$ is a flat $\mathcal{O}_X$-module). φ and b are both injective so a is injective. Hence F is torsion-free as well as torsion. Hence $F = 0$ so $\varphi = 0$.

CHAPTER THREE

QUASICOHERENT SHEAVES ON NOETHERIAN LOCALLY SEPARATED ALGEBRAIC SPACES

1. The Completeness/Extension Lemma 153

2. The Serre Criterion 159

3. Schemehood and Nilpotents 165

4. Chevalley's Theorem 169

5. Devissage 173

1. The Completeness/Extension Lemma

Let X be a noetherian locally separated algebraic space. The Completeness/Extension Lemma is a basic tool for relating the categories of quasicoherent and coherent sheaves on X. There are two assertions involved. Completeness says that every quasicoherent sheaf on X is the union of its coherent subsheaves. Phrased in the terminology of Gabriel XIV

the category of quasicoherent sheaves on a locally separated noetherian algebraic space is a "locally noetherian category."

The Extension assertion is that, for every open immersion $i : U \to X$, and every quasicoherent sheaf F on X, and for every coherent subsheaf G_U of the restriction i^*F of F to U, there is a coherent subsheaf G of F on X with $i^*G = G_U$.

These are equivalent assertions in the category of schemes in the sense that each can be proved easily from the other but the proof of either from scratch is more involved. The difficulty involved can be seen in the extension assertion: the required coherent sheaf G is not canonically determined by G_U but must be constructed ad hoc.

The idea of the proof in the case of schemes is as follows: We first observe that completeness is trivial for affine schemes. Hence the Extension lemma holds for open subschemes U of affine schemes X. Now let X be any scheme and $\{U_i\}_{i \in I}$ a covering of X by affine open subschemes U_i. First assume $I = \{1,2\}$, so $X = U_1 \cup U_2$. Let F be a quasicoherent sheaf on X, and let G_1 be any coherent subsheaf of $F\big|_{U_1}$. Then $G_1\big|_{U_1 \cap U_2}$ is a coherent subsheaf of $F\big|_{U_1 \cap U_2}$ so by the Extension lemma for $U_1 \cap U_2 \subset U_2$, there is a coherent sheaf G_2 of $F\big|_{U_2}$ with $G_2\big|_{U_1 \cap U_2} = G_1\big|_{U_1 \cap U_2}$. Hence G_1 and G_2 glue together to give a coherent subsheaf G of F on X with $G\big|_{U_1} = G_1$. This shows that if $\overline{F}$ is the union of the coherent subsheaves of F, then $\overline{F}\big|_{U_1} = F\big|_{U_1}$. Similarly $\overline{F}\big|_{U_2} = F\big|_{U_2}$. Hence $\overline{F} = F$. Thus any scheme which is covered by at most two affine schemes satisfies completeness, hence satisfies the extension lemma. One proceeds to the general case where X is covered by a finite number of affines by induction.

Unfortunately such a technique is not available to us since the topology on an algebraic space is given by affine schemes U_i mapping etale to X, and not necessarily injectively. (And the extension lemma is obviously false for an arbitrary etale map $U \longrightarrow X$ - just take U to be the disjoint union of two copies of X!) Hence we prove the completeness assertion in a different way and deduce the extension assertion as a corollary.

Our main reason for proving this lemma is its use in the proof of Devissage (5.2) and in the proof of the Serre Criterion (2.5).

<u>Theorem 1.1</u> (The Completeness Lemma.) Let F be a quasicoherent sheaf on a noetherian algebraic space X. Then $F = \varinjlim F_i$, the limit being taken over all coherent subsheaves F_i of F.

<u>Proof:</u> (P. Deligne) By II.2.9, we can find an etale covering of X, $\pi : U \longrightarrow X$, with U a noetherian affine scheme, and π quasicompact and separated. The completeness assertion is certainly true for

affine algebraic spaces, so π^*F is a union of coherent sub - $\mathcal{O}_U$-modules $\overline{F}_i$.

π_* is not right exact (unless X is separated) but it does preserve unions (in fact all filtered direct limits - see, e.g., GT II.5.4). Hence $\pi_*\pi^*F = \varinjlim \pi_*\overline{F}_i$.

We now define F_i as the pullback (intersection) in the following diagram:

$$\begin{array}{ccc}
F_i & \hookrightarrow & \pi_*\overline{F}_i \\
\downarrow & & \downarrow \\
F & \hookrightarrow & \pi_*\pi^*F
\end{array}$$

By the AB_5 property (II.4.3), $F = \varinjlim F_i$.

Finally, for each i, there is a commutative diagram of $\mathcal{O}_U$-modules:

$$\begin{array}{ccccc}
\pi^*F_i & \longrightarrow & \pi^*\pi_*\overline{F}_i & \longrightarrow & \overline{F}_i \\
& \searrow & & \swarrow & \\
& & \pi^*F & &
\end{array}$$

Since F_i is a subsheaf of F, and is flat, the map $\pi^*F_i \to \pi^*F$ is injective. Hence the top map $\pi^*F_i \to \overline{F}_i$ is injective. $\overline{F}_i$ is coherent so π^*F_i is coherent. π is a covering so F_i is coherent. ∎

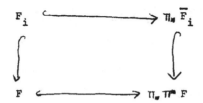

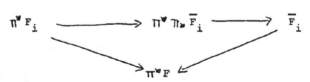

<u>Corollary 1.2</u> (The Extension Lemma). Let $U \to X$ be an open subspace of a locally separated noetherian algebraic space X. Let F be a quasicoherent sheaf on X, and $G_{\mathcal{U}} \to i^*F$ a coherent sub-$\mathcal{O}_{\mathcal{U}}$-module of i^*F. Then there is a coherent sheaf G on X with $G_{\mathcal{U}} = i^*G$.

<u>Proof</u>: We define G_1 by the cartesian diagram of $\mathcal{O}_X$-modules

$$\begin{array}{ccc} G_1 & \longrightarrow & F \\ \downarrow & & \downarrow \\ i_*G_u & \longrightarrow & i_*i^*F \end{array}$$

The bottom map is injective so G_1 is a subsheaf of F.

Apply i^*, which is exact, to this diagram. Note $i^*i_*Q = Q$ for all quasicoherent Q on U since $U \to X$ is injective. Thus

$$\begin{array}{ccc} i^*G_1 & \longrightarrow & i^*F \\ \downarrow & & \downarrow \\ G_{\mathcal{U}} & \longrightarrow & i^*F \end{array} \quad \text{is cartesian,}$$

so $i^*G_1 = G_{\mathcal{U}}$.

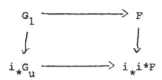

Hence there is some subsheaf of F which pulls back by i^* to $G_{\mathcal{U}}$. Since G is coherent, and $G_1 = \varinjlim_{\substack{G_i \to G_1 \\ G_i \text{ coherent}}} G_i$ by the lemma, there is a coherent $G \to G_1$ with $i^*G = G_{\mathcal{U}}$. ∎

1.3. It should be remarked here that the proof of the Completeness Theorem is much easier in the case of separated noetherian algebraic spaces X. In this case, let $f: U \to X$ be an etale surjective map with U affine. Then f is affine. One first remarks the lemma B above. Then for any quasicoherent sheaf F on X, there is an immersion $F \to f_*f^*F$ so it is sufficient to prove the theorem for sheaves of the form f_*G, G quasicoherent on U. Now we copy the statement and proof of lemma D, noting that the only fact used there is that the map $U \to X$ is affine.

<u>Corollary 1.4</u>. Let X be an integral noetherian locally separated algebraic space. Let F be a coherent sheaf on X. Then there is an integer m, and a coherent sheaf H on X and maps $\alpha : H \to \mathcal{O}_X^m$ and $\beta : H \to F$ such that for some dense open subspace U of X, the restrictions of α and β to U are isomorphisms.

<u>Proof</u>: By II.8.13 there is a dense open subspace U of X such

that the restriction of F to U is a free $\mathcal{O}_U$-module. Thus there is an isomorphism $\overline{\Pi} : \mathcal{O}_U^m \to F|_U$ of $\mathcal{O}_U$-modules. Let $H_U \subset (\mathcal{O}_X^m \oplus F)|_U$ be the graph of this isomorphism. By the extension lemma, there is a coherent subsheaf H of $\mathcal{O}_X^m \oplus F$ on X with $H|_U = H_U$. The projection maps then give two maps $H \to \mathcal{O}_X^m$ and $H \to F$. Each of these maps is an isomorphism on U.∎

2. The Serre Criterion

The Serre Criterion is a cohomological criterion for a separated noetherian algebraic space to be a scheme. As in the case of the Completeness Lemma of section 1, the proof for schemes, as given in EGA, does not generalize directly to the category of algebraic spaces.

In producing a new proof for this theorem, we found that there are apparently two assertions involved. The first, which we label the Weak Serre Criterion, holds for quasicompact separated algebraic spaces and is a purely formal abelian category and descent theory fact. Several applications of the criterion (e.g., the results in section 3) are consequences of just this weak statement.

The full Serre Criterion follows from the weak Serre Criterion and the Completeness Lemma. Thus we have this theorem only in the case of noetherian separated algebraic spaces.

(This is slightly weaker than the Serre Criterion for schemes which holds for arbitrary quasicompact separated schemes.)

Before stating and proving this theorem, we review some of the theory of abelian categories.

Notation 2.1. Let $\mathcal{A}$ and $\mathcal{B}$ be abelian categories and $F : \mathcal{A} \to$ an additive functor. F is exact if for every short exact sequence $0 \to M' \to M \to M'' \to 0$ in $\mathcal{A}$, the sequence $0 \to F(M') \to F(M) \to F(M'') \to 0$ is exact in $\mathcal{B}$. F is faithful if for every M in $\mathcal{A}$, $M \neq 0$, $F(M) \neq 0$. (If F is exact this is equivalent to the assertion that for every M, N the map $\mathrm{Hom}_{\mathcal{A}}(M,N) \to \mathrm{Hom}_{\mathcal{B}}(F(M),F(N))$ is injective.)

Now let $\mathcal{A}$ be an abelian category and (Ab) the category of abelian groups. We assume $\mathcal{A}$ has arbitrary sums (= coproducts). An object P in $\mathcal{A}$ is faithfully projective if the functor $\mathrm{Hom}_{\mathcal{A}}(P,-) : \mathcal{A} \to (\mathrm{Ab})$ is exact, faithful, and preserves sums. Let A be the ring $\mathrm{Hom}_{\mathcal{A}}(P,P)$. Then for every $M \in \mathcal{A}$, $\mathrm{Hom}\ (P,M)$ is an A-module so $\mathrm{Hom}_{\mathcal{A}}(P,-)$ gives a functor $F : \mathcal{A} \to$ (A-modules), $F(M) = \mathrm{Hom}_{\mathcal{A}}(P,M)$.

Theorem 2.2. (Gabriel, Mitchell) With the assumptions of 2.1, if P is faithfully projective, F is an equivalence of categories.

Proof: See Bass, Algebraic K-Theory, p. 54. ∎

We are now ready to prove our theorem. As usual, QCS_X denotes the category of quasicoherent sheaves on an algebraic space X, and $\mathcal{O}_X$ the structure sheaf of X.

Theorem 2.3. (The Weak Serre Criterion) Let X be a quasicompact separated algebraic space. Suppose the global section functor $\Gamma(X,-) : QCS_X \to (Ab)$ is exact and faithful. Then X is an affine scheme.

Proof: Since X is quasicompact, QCS_X has all sums and $\Gamma(X,-)$ commutes with direct sums. (See II.4.17, and [GT]). For any $F \in QCS_X$, $\Gamma(X,F) = \mathrm{Hom}_{QCS_X}(\mathcal{O}_X, F)$. Thus the hypotheses say that $\mathcal{O}_X$ is a faithfully projective object in QCS_X, so by the theorem 2.2, $\Gamma(X,-) : QCS_X \to$ $(\Gamma(X, \mathcal{O}_X)$-modules) is an equivalence of categories.

Let $A = \Gamma(X, \mathcal{O}_X) = \mathrm{Hom}_{QCS_X}(\mathcal{O}_X, \mathcal{O}_X)$. The category of A-modules is equivalent to the category $QCS_{\mathrm{Spec}\ A}$, so we have a functor $\Gamma : QCS_X \to QCS_{\mathrm{Spec}\ A}$ which is an equivalence of categories.

We now claim that there is a natural map of algebraic spaces $\gamma : X \to \mathrm{Spec}\ A$ such that γ is quasicompact and separated. Applying II.4.6, $\gamma_* : QCS_X \to QCS_{\mathrm{Spec}\ A}$ and we claim $\gamma_* = \Gamma$.

To see this, let Spec B → X be an affine etale covering of X. Since X is separated, Spec B $\times_X$ Spec B is affine--call this scheme Spec C. Then the exact sequence of rings $A \to B \rightrightarrows C$ induces a map of algebraic spaces

$\mathrm{Cok}_{\mathrm{Alg.Spaces}}(\mathrm{Spec}\ C \rightrightarrows \mathrm{Spec}\ B) \xrightarrow{\gamma} \mathrm{Spec}\ A$. Since X is quasicompact and separated, and Spec A is affine, γ is quasicompact and separated. It is then clear that $\gamma_* = \Gamma$.

γ_* is an equivalence of categories, and its inverse equivalence $\gamma_*^{-1} : \mathrm{QCS}_{\mathrm{Spec}\ A} \to \mathrm{QCS}_X$ must be a left adjoint of γ_*. But γ^*, the usual inverse image of quasicoherent sheaves, is a left adjoint of γ_*. By uniqueness of the adjoint, γ^* is naturally equivalent to γ^{-1}. From this we point two consequences: first, $\gamma^*\gamma_* : \mathrm{QCS}_X \to \mathrm{QCS}_X$ is naturally equivalent to the identity functor, and second, γ^* is exact and faithful, so the map $\gamma : X \to \mathrm{Spec}\ A$ is faithfully flat.

We now apply the Sheaf Criterion for isomorphism (II.5.3) which says that γ must be an isomorphism, so X is affine. ∎

Lemma 2.4: Let X be a noetherian algebraic space and F be a coherent sheaf on X. Let I be a sheaf of ideals on X. Suppose IF = F. Then $\mathrm{Supp}(F) \cap V(I) = \emptyset$.

<u>Proof:</u> The conditions are local on X so we can assume X is affine. Hence we have a ring R, ideal I, module M of finite type, and IM = M. Let p be any prime ideal representing a point $x_p \in X = \text{Spec } R$, such that $x_p \in V(I)$. Then $p \supseteq I$ so $pM = M$. Hence by the Nakayama lemma, the associated sheaf $\tilde{M}$ is zero in a neighborhood of x_p, so $x_p \notin \text{Supp } M$.

<u>Theorem 2.5:</u> (The Serre Criterion) Let X be a separated noetherian algebraic space. Suppose the global section functor $\Gamma(X,-)$: (Coherent Sheaves on X) $\to$ (Ab) is exact. Then X is affine.

<u>Proof:</u> We use the hypothesis in the form: $H^1(X,F) = 0$ for all coherent sheaves F on X.

For any quasicoherent sheaf F on X, the Completeness Lemma says that $F = \varinjlim F_i$ the (filtered) direct limit of all coherent subsheaves. By II.4.17, $H^1(X,F) = \varinjlim H^1(F,F_i)$ so $H^1(X,F) = 0$ for all quasicoherent sheaves F on X. Thus, using the Weak Serre Criterion, we can conlude that X is affine by just showing that $\Gamma(X,-)$ is also faithful. To show this, we use noetherian induction and assume that every closed subspace X_1 of X, with $X_{1red} \neq X_{red}$, is affine.

The induction hypothesis can be applied to coherent sheaves on X in the following way. Let F be a coherent sheaf on X such that $\Gamma(X,F) = 0$ and the support of F is a proper closed subset of X. By II.5.14 there is a proper closed subspace $i : C \to X$ such that $F = i_*F_1$ for some F_1 on C. Then $\Gamma(X,F) = \Gamma(X,i_*F_1) = \Gamma(C,F_1) = 0$ and since C is affine $F_1 = 0$. Hence $F = 0$.

Let $U \to X$ be an affine open subspace. If $U = X$ we are done; X is affine. Suppose not. Let I be a sheaf of ideals such that $\operatorname{Spec} \mathcal{O}_{X/I} = X - U$. Let F be a coherent sheaf on X with $\Gamma(X,F) = 0$. Consider the exact sequence

$$0 \to IF \to F \to F/_{IF} \to 0$$

Applying the (exact) global section functor

$$0 \to \Gamma(X,IF) \to 0 \to \Gamma(X,F/_{IF}) \to 0$$

Hence $\Gamma(X,F/_{IF}) = 0$. Since the support of $F/_{IF}$ is a proper closed subset of X, (U is nonempty), $F/_{IF} = 0$ by the induction hypothesis. Hence $IF = F$. Using lemma 2.4, $\operatorname{Supp} F \cap C = \emptyset$. Since C is nonempty, $\operatorname{Supp} F$ is a proper closed subset of X so $F = 0$.

Now let F be any quasicoherent sheaf on X such that $\Gamma(X,F) = 0$. By the Completeness Lemma, F is the union of its coherent subsheaves, $F = \bigcup F_i$. Then $0 = \Gamma(X,F) = \bigcup \Gamma(X,F_i)$ so each $\Gamma(X,F_i) = 0$. Hence each $F_i = 0$ so $F = 0$. ∎

3. Schemehood and Nilpotents

Lemma 3.1. Let $f : X \to Y$ be an immersion of algebraic spaces. Then f is a closed immersion if and only if the associated map of reduced algebraic spaces $f_{red} : X_{red} \to Y_{red}$ is a closed immersion.

Proof: The question is local on Y so we can take Y to be an affine scheme, in which case X is also a scheme and in the category of schemes, this is clear. ∎

Lemma 3.2. Let X be a locally separated algebraic space. Then X is separated if and only if X_{red} is separated.

Proof: Note first that $(X \times X)_{red} = (X_{red} \times X_{red})_{red}$ since the second satisfies the universal property for $(X \times X)_{red} \to X \times X$. In the commutative triangle

$$\begin{array}{ccc} X_{red} & \xrightarrow{\quad i \quad} & (X \times X)_{red} = (X_{red} \times X_{red})_{red} \\ & {\scriptstyle\Delta} \searrow \quad \swarrow {\scriptstyle j} & \\ & X_{red} \times X_{red} & \end{array}$$

j is a closed immersion, so separated. Applying I.1.21,

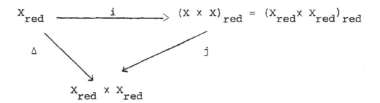

Δ is a closed immersion if and only if i is a closed immersion, which by Lemma 3.1, happens if and only if $X \to X \times X$ is a closed immersion. ∎

<u>Theorem 3.3</u>. Let X be a noetherian locally separated algebraic space. Then X is an affine scheme if and only if X_{red} is an affine scheme.

<u>Proof</u>: By Lemma 3.2, we can assume X is separated. If X is affine, then clearly the closed subspace X_{red} of X is affine.

For the converse, let J be the sheaf of nilpotents of X, so $X_{red} = \underset{\sim}{\mathrm{Spec}}\, \mathcal{O}_{X/J}$. Since X is noetherian, J is coherent so there is an integer n with $J^n = 0$. Let $X_i = \underset{\sim}{\mathrm{Spec}}\, \mathcal{O}_{X/_{J^i}}$ so that there is a sequence of closed immersions $X_{red} = X_1 \to X_2 \to \ldots \to X_n = X$. Each map $X_i \to X_{i+1}$ is a closed immersion with $X_i = \underset{\sim}{\mathrm{Spec}}\, \mathcal{O}_{X_{i+1/I}}$ with $I = J^i/_{J^{i+1}}$ an ideal satisfying $I^2 = 0$. Hence we are reduced to the following lemma:

<u>Lemma</u>. Let $f : Y \to Z$ be a closed immersion of noetherian separated algebraic spaces with $Y = \mathrm{Spec}\, \mathcal{O}_{Z/I}$, $I^2 = 0$. Then if Y is affine, so is Z.

Proof of Lemma. Let F be a quasicoherent sheaf on Z. Consider the exact sequence $0 \to IF \to F \to F/_{IF} \to 0$. IF and $F/_{IF}$ are both annihilated by I so they are $\mathcal{O}_{Z/I} = \mathcal{O}_Y$-modules. Apply the global section functor $\Gamma(Z,-)$:

$$0 \to \Gamma(Z,IF) \to \Gamma(Z,F) \to \Gamma(Z,F/_{IF})$$

$$H^1(Z,IF) \to H^1(Z,F) \to H^1(Z,F/_{IF})$$

Since IF and $F/_{IF}$ are $\mathcal{O}_Y$-modules, $H^1(Z,IF) = H^1(Y,IF)$ and $H^1(Z,F/_{IF}) = H^1(Y,F/_{IF})$. Y is affine so both of these vanish. Hence $H^1(Z,F) = 0$ so $\Gamma(Z,-)$ is exact. By the Serre Criterion, Z is affine. ∎

Proposition 3.4. Let X be a noetherian locally separated algebraic space. Then there is a one-one correspondence between separated etale maps $Y \to X$ and separated etale maps $Y' \to X_{red}$ given by

$$Y \to X \rightsquigarrow Y' = Y \underset{X}{\times} (X_{red}) \to X_{red}.$$

This correspondence preserves and reflects affine schemes.

Proof: The one-one correspondence is local on X so we can assume that X is affine. Hence for any separated etale map $Y \to X$, Y must be a scheme (by II.6.5). In the category of schemes, the one-one correspondence is I.4.20.

Note that $Y \times_X (X_{red}) = Y_{red}$. By 3.3, Y is affine if and only if Y_{red} is affine. ∎

Corollary 3.5. Let X be a noetherian locally separated algebraic space and $f : X_{red} \to X$ the canonical inclusion. Then the two functors

$$f_* : AS_{X_{red}} \to AS_X \qquad f^* : AS_X \to AS_{X_{red}}$$

are inverses to each other. (AS as usual here means the category of abelian sheaves.) ∎

Corollary 3.6. Let X be a noetherian locally separated algebraic space. Then X is a scheme if and only if X_{red} is a scheme.

Proof: An open covering of X_{red} by affine schemes lifts to an open covering of X by affine schemes. ∎

4. Chevalley's Theorem

This theorem gives another criterion for a noetherian separated algebraic space to be a scheme. Its application for us will be in the proof of Chow's Lemma (See IV.3.1).

Theorem 4.1: (Chevalley's Theorem) Let X be an affine scheme and Y a noetherian separated algebraic space. Suppose there is a map $f : X \to Y$ which is finite and surjective. Then Y is an affine scheme.

Proof: Consider $X \times_Y Y_{red} \to Y_{red}$, also a finite surjective morphism. By Corollary 3.3 or the Serre Criterion, it is sufficient to show that Y_{red} is affine. $X_{red} \to X \times_Y Y_{red}$ is finite and surjective. Thus we can assume that X and Y are reduced.

We prove the theorem by noetherian induction on Y, assuming that for every closed subspace Y_1 of Y, with $Y_{1red} \neq Y_{red}$, Y_1 is affine. Hence for every coherent sheaf F on Y, with $\mathrm{Supp}\, F \neq Y$, $H^1(Y,F) = 0$.

Suppose Y is not irreducible. Let $Y' \xrightarrow{j} Y$ be a component of Y. For any coherent $\mathcal{O}_Y$-module, F, let $F' = j_* j^* F$ and $\rho : F \to F'$ be the natural map. Let $G = \mathrm{Ker}\, \rho$

and $K = \text{Im}\,\rho$. ρ is an isomorphism on $Y' - (Y \cap Y')$ so G and K have support not equal to Y. Hence $H^1(Y,G) = H^1(Y,K) = 0$ so $H^1(Y,F) = 0$. This holds for all coherent sheaves F on Y so by the Serre Criterion 2.5, we are done; Y is affine.

On the other hand, suppose Y is irreducible, so integral. We now need a lemma:

Lemma: In the situation above, there is an integer m, and a map of sheaves $u : \mathcal{O}_Y^m \to f_*\mathcal{O}_X$ such that on a dense affine open subspace U of Y, the restriction of u to U is an isomorphism.

Proof of lemma: The map f is finite so $f_*\mathcal{O}_X$ is a coherent $\mathcal{O}_Y$-module. Applying II. 8.13 , there is a dense affine open subspace U of Y and an isomorphism $u_1 : \mathcal{O}_Y^m|_U \to f_*\mathcal{O}_X|_U$ for some integer m. Let $U' = f^{-1}(U) = U \times_Y X$. Let $s_1,\ldots,s_m \in \Gamma(U',\mathcal{O}_X) = \Gamma(U,f_*\mathcal{O}_X)$ be the elements defining u_1. X is an affine scheme so there is an element $g \in \Gamma(X,\mathcal{O}_X)$, with the restriction $g' \in \Gamma(U',\mathcal{O}_X)$ a unit, and with $g's_1,\ldots,g's_m$ extendable to sections $b_1,\ldots,b_m$ of $\Gamma(X,\mathcal{O}_X)$. These $\{b_i\}$ define $u : \mathcal{O}_Y^m \to f_*\mathcal{O}_X$ and $u|_U : \mathcal{O}_Y^m|_U \to f_*\mathcal{O}_X|_U$ is the map taking the element

$(0,\ldots,1,\ldots,0)$ with 1 in the ith position to $b_i|_U = g's_i$. But since g' is a unit on U', there is an isomorphism $w : f_*\mathcal{O}_X|_U \to f_*\mathcal{O}_X|_U$ taking any $x \rightsquigarrow x/g'$. The composite $w\cdot u|_U$ is the isomorphism u_1. Hence $u|_U$ is an isomorphism.

We can now prove the theorem. Our induction assumption says that every torsion sheaf F on Y satisfies $H^1(Y,F) = 0$. By II.8.15, every sheaf F on Y fits into an exact sequence $0 \to F' \to F \to F'' \to 0$ with F' torsion and F'' torsion-free. Hence it is sufficient to show that for every torsion-free sheaf F, $H^1(Y,F) = 0$.

Let F be a torsion-free sheaf on Y and $G = \mathrm{Hom}(f_*\mathcal{O}_X, F)$. Then the map $u : \mathcal{O}_Y^m \to f_*\mathcal{O}_X$ constructed above defines a map $v : G = \mathrm{Hom}(f_*\mathcal{O}_X, F) \to \mathrm{Hom}(\mathcal{O}_Y^m, F) = F^m$. The exact sequence $\mathcal{O}_Y^m \to f_*\mathcal{O}_X \to T \to 0$, where T is the cokernel, necessarily a torsion sheaf, gives an exact sequence $0 \to \mathrm{Hom}(T,F) \to G \overset{v}{\to} F^m$. T is torsion and F is torsion-free so by II.8.15, $\mathrm{Hom}(T,F) = 0$. Hence v is injective.

Now consider the exact sequence $0 \to G \overset{v}{\to} F^m \to \mathrm{Cok}\ v \to 0$. $\mathrm{Cok}\ v$ is torsion so $H^1(Y, \mathrm{Cok}\ v) = 0$. $G = \mathrm{Hom}(f_*\mathcal{O}_X, F)$ is an $\mathcal{O}_X$-module. $X \to Y$ is an affine map. Hence there is a coherent sheaf G_1 on X with $G = f_*G_1$. Hence $H^1(Y,G) = H^1(X,G_1) = 0$, so $H^1(Y,F) = 0$. This holds for all torsion-free sheaves on Y, hence for all coherent sheaves on Y. By the Serre Criterion, Y is affine. ∎

Corollary 4.2. Let

$$\begin{array}{ccc} X & \xrightarrow{f} & Y \\ & {\scriptstyle h}\searrow \quad \swarrow{\scriptstyle g} & \\ & S & \end{array}$$

be a commutative diagram of noetherian algebraic spaces with h an affine map, g separated and of finite type, and f surjective and finite. Then g is affine.

Proof: The assertion is local on S so we can assume that S is an affine scheme in which case this is exactly the Chevalley Theorem. ∎

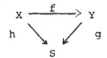

5. Devissage

Devissage is an inductive technique for the category of coherent sheaves on a noetherian locally separated algebraic space. We will use this technique in Chapter IV for the proof of the finiteness theorem.

The proof here is a translation of the equivalent assertion in EGA (III.2.1.2). (We note that the definition of exact in 5.1 is slightly altered - Grothendieck does not assume the direct sum condition but uses it as an extra hypothesis in the theorem.)

Definition 5.1. Let K be an abelian category. A subset K' of the objects of K is exact if

1) $0 \in K'$
2) For any exact sequence $0 \to A' \to A \to A'' \to 0$ in K, if two terms are in K', so is the third.
3) If $A_1 \oplus A_2 \in K'$, then A_1 and A_2 are in K'.

Theorem 5.2. (Devissage) Let X be a noetherian locally separated algebraic space and K the category of all coherent $\mathcal{O}_X$-modules. Let K' be an exact subcategory of K such that for every integral closed subspace $Y \to X$, there is a sheaf $G \in K'$ with $\operatorname{Supp} G = Y$. Then every coherent sheaf belongs to K'. (Recall Supp G = Y for a coherent sheaf G and integral algebraic space Y entails that the stalk G_y of G at the generic point y of Y is nonzero.)

Proof: Consider the following property P(Y) of a closed subscheme Y of X: Every coherent $\mathcal{O}_X$-module with support contained in Y belongs to K'. By noetherian induction it remains to show that if Y is a closed subscheme of X such that P(Y') holds for every closed subscheme Y' of Y, then P(Y) is true.

Let $F \in K$ have support contained in Y. We will prove $F \in K'$. Say Y_{red} is defined by an ideal I of $\mathcal{O}_X$. By (II.5.8) there is an integer n such that $I^nF = 0$. Then there are exact sequences

$$0 \to I^{n-1}F/_{I^nF} \to F/_{I^nF} \to F/_{I^{n-1}F} \to 0$$

of coherent $\mathcal{O}_X$-modules. Since K' is exact, it is sufficient by induction to show that each $F_K = I^{k-1}F/_{I^kF}$ is in K'. Thus we can assume $IF = 0$, i.e. that $F = j_*(j^*(F))$ where j is the injection $Y_{red} \to X$. Thus we can assume Y is reduced. We now distinguish two cases:

a) Y is reducible. Say $Y = Y' \cup Y''$ with Y' and

Y" two reduced closed subspaces of Y distinct from Y. Say Y', Y" are defined by ideals J', J" of $\mathcal{O}_Y$. Put

$$F' = F \otimes_{\mathcal{O}_Y} (\mathcal{O}_{Y/J'}) \quad \text{and} \quad F'' = F \otimes_{\mathcal{O}_Y} (\mathcal{O}_{Y/J''}).$$

The canonical maps $F \to F'$ and $F \to F''$ define a map $a : F \to F' \oplus F''$. a is an isomorphism on $Y - (Y' \cap Y'')$ so Ker a and Cok a are in K', using the induction hypothesis. Hence $F \in K'$ iff Im $a \in K'$ iff $F' \oplus F'' \in K'$ iff F' and F" are each in K'. Since F' and F" each have support unequal to Y, we are done.

b) Y is irreducible, so integral. Let F be any coherent sheaf on Y. Applying 1.4, there is an integer m and a coherent $\mathcal{O}_Y$-module H and maps $\alpha : H \to \mathcal{O}_Y^m$ and $\beta : H \to F$ such that on some dense open subspace U of Y, the restrictions of α and β are isomorphisms. Thus Ker α, Cok α, Ker β, and Cok β, all have support not equal to Y. By the induction assumption, they are all in K'. Hence $F \in K'$ iff Im $\beta \in K'$ iff $H \in K'$ iff Im $\alpha \in K'$ iff $\mathcal{O}_Y^m \in K'$.

We apply this first to the postulated coherent sheaf G for which supp G = Y. This shows that $\mathcal{O}_Y^m \in K'$ for some $m \neq 0$. Hence $\mathcal{O}_Y^n \in K'$ for any integer $n \geq 0$. Now applying the argument for any coherent sheaf F on Y, $\mathcal{O}_Y \in K'$ so $F \in K'$. ∎

CHAPTER FOUR

THE FINITENESS THEOREM

1. Actions of a Finite Group 177

2. Symmetric Powers of Projective Spaces...... 185

3. Chow's Lemma 192

4. The Finiteness Theorem 202

As indicated in the title, the major theorem in this chapter is the Finiteness Theorem. The proof is the same as that in the case of schemes (cf. EGA III.3.2.1) where one applies an inductive procedure, specifically Devissage, using Chow's Lemma for the initial step. The problem in the generalization to algebraic spaces lies not in the final proof of Finiteness but rather in the generalizations of the ingredients. Chapter III can be viewed as essentially a proof of Devissage; this chapter as essentially a proof of Chow's Lemma.

The proof of Chow's Lemma uses the notion of symmetric powers of projective spaces, and a few notions about finite groups acting on algebraic spaces. For the latter, we give in Section 1 the bare minimum of foundations necessary for the later proof. The reader hopeful of finding the general foundations of the theory of finite group actions on algebraic spaces should see 1.8.

In this chapter, all algebraic spaces are assumed to be *Noetherian* and *separated*.

1. *Actions of a Finite Group*

Definition 1.1: Let G be a finite group. A *G-space* is an algebraic space X with a group homomorphism $G \to \mathrm{Aut}(X)$. A *G-map* of G-spaces $f:X \to Y$ satisfies $fq = qf$ for all q in G. A G-space is *trivial* if every q in G acts as the identity map.

Let X be a G-space and Y be any algebraic space. A *G-invariant* map is a G-map $f:X \to Y$ where Y is taken as a trivial G-space. (Note in this case there is a map of sheaves $\mathcal{O}_Y \to (f_* \mathcal{O}_X)^G$ where the superscript "G" denotes as usual the fixed elements under the action of G.)

Let E and X be G-spaces. A *stable open subspace* (resp. *stable etale map*) $f:E \to X$ is a G-map which is an open immersion (resp. an etale map) and which satisfies, for all q in G,

$$\begin{array}{ccc} E & \xrightarrow{q} & E \\ \downarrow f & & \downarrow f \\ X & \xrightarrow{q} & X \end{array}$$

is cartesian.

Let X be a G-space. A map of algebraic spaces $\pi: X \to Y$ is a <u>geometric quotient of X by G</u> if

1) π is G-invariant and finite

2) The map of sheaves $\mathcal{O}_Y \to (\pi_* \mathcal{O}_X)^G$ is an isomorphism.

Note that, as an immediate consequence, π is affine and proper. Also, by Chevalley's Theorem, X is affine if and only if Y is affine.

<u>Example 1.2</u>: Let X be an algebraic space and G a finite group acting on X. G acts <u>freely</u> if for any $\sigma \in G$, the fixed point locus of σ is empty. I.e., the following diagram is cartesian:

$$\begin{array}{ccc} \text{empty space} & \longrightarrow & X \\ \downarrow & & \downarrow {\scriptstyle 1 \times \sigma} \\ X & \xrightarrow{\Delta} & X \times X \end{array}$$

We write $G \times X$ for the disjoint sum of copies of X, one for each $\sigma \in G$, and use the notation $(\sigma, x) \in G \times X$ to mean the point x in the σth copy of X. There are two maps

$$G \times X \underset{\pi_2}{\overset{\pi_1}{\rightrightarrows}} X$$

defined by $\pi_1(\sigma, x) = x$, $\pi_2(\sigma, x) = \sigma(x)$, and this defines an etale equivalence relation on X, whose quotient $\pi: X \to Y$ is a geometric quotient.

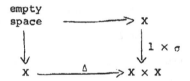

<u>Proposition 1.3</u>: Let X be a G-space and $\pi : X \to Y$ a geometric quotient. Let $Y' \to Y$ be an etale map. Then under the natural action of G on $Y' \times_Y X$, $Y' \times_Y X \to Y'$ is a geometric quotient. Conversely, if G acts on X, and $X \to Y$ is G-invariant, $Y' \to Y$ is etale surjective, and $Y' \times_Y X \to Y'$ is a geometric quotient, then so also is $X \to Y$.

<u>Proof</u>: The definition of geometric quotient is clearly local on Y. ∎

<u>Proposition 1.4</u>: Let X be a G-space and $\pi : X \to Y$ a geometric quotient. Let $Y' = \operatorname{Spec} \mathcal{O}_Y[Z_1,\ldots,Z_n]$ be affine n-space over Y. Let $X' = Y' \times_Y X$, affine n-space over X, be a G-space under the induced action. Then $X' \to Y'$ is a geometric quotient.

<u>Proof</u>: $X' \to Y'$ is clearly G-invariant and finite. To check the quotient property, we can use the last proposition to reduce to the affine case. Thus if $X = \operatorname{Spec} A$ and $Y = \operatorname{Spec} A^G$, we have

$$\begin{array}{ccc} A[Z_1,\ldots,Z_n] & \longleftarrow & A \\ \uparrow & & \uparrow \\ A^G[Z_1,\ldots,Z_n] & \longleftarrow & A^G \end{array}$$

and the assertion is that

$$A^G[Z_1,\ldots,Z_n] = (A[Z_1,\ldots,Z_n])^G$$

which is clear. ∎

<u>Proposition 1.5</u>: Let $X \to S$ be a quasiprojective map of algebraic spaces. Suppose X is a G-space where each $\sigma \in G$ acts as an S-automorphism. Then there is a unique S-space Y, and map $\pi : X \to Y$ which is a geometric quotient of X by G. Y is also quasiprojective over S. Furthermore, if $U \to X$ is a stable open subspace of X, and $\pi' : U \to V$ is a geometric quotient of U, then there is a cartesian diagram

$$\begin{array}{ccc} U & \longrightarrow & X \\ {\scriptstyle \pi'}\downarrow & & \downarrow{\scriptstyle \pi} \\ V & \longrightarrow & Y \end{array}$$

of S-spaces, where $V \to Y$ is an open immersion.

<u>Proof</u>: The proposition is local on S, so we can take S to be affine, in which case all of this except the quasiprojectivity of Y is quoted from SGA 60-61 V.

We indicate the proof of quasiprojectivity. Let $f : Y \to P^m_S$ be a projective imbedding of Y, and consider the composites $Y \xrightarrow{\sigma} Y \xrightarrow{f} P^m_S$, $\sigma \in G$. These define a map $Y \to (P^m)^n$ (where n is the order of G). Imbed $(P^m)^n$ into $P^{(m+1)^n-1}$ by the Segre map (EGA II. 4.3). We then have a projective imbedding of Y such that the group G acts linearly on Y. Let $\bar{Y}$ be the closure of Y in $P^{(m+1)^n-1}$. The action of G extends to $\bar{Y}$ and $Y/G \to \bar{Y}/G$ is an open immersion. Hence we can assume $Y = \bar{Y}$. Thus $Y = \operatorname{Proj} A$, with A a graded S-algebra on which G acts linearly.

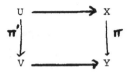

The geometric quotient Spec $A \rightarrow$ Spec A^G gives by restriction a map Proj $A \rightarrow$ Proj A^G (here the linearity of the action is essential.) The only difficulty now is to show that Proj A^G is projective. Since A is of finite type over S, S is noetherian, and G is finite, A^G is of finite type, so has a finite number of generaters. Unfortunately these generaters will not all be in the degree 1 part of $A^G = \bigoplus A_n^G$ (as they need be to apply EGA II.5.5.1 and hence to assert that Proj A^G is projective). Instead the generaters lie in degrees between 1 and some integer k. We get around this difficulty by considering $(A^G)^{(k!)} = \bigoplus_{r=0}^{\infty} A_{rk!}^G$ for which (by EGA II.2.4.7) Proj $(A^G)^{(k!)} =$ Proj A^G. A bit of combinatorics shows this algebra has the property that it is generated over A_0^G by the terms of degree 1.∎

Proposition 1.6: Let X be a G-space and $\pi: X \rightarrow Y$ be a geometric quotient. Then for every $Y' \rightarrow Y$ of finite type, $\pi': X \times_Y Y' \rightarrow Y'$ is an open morphism.

Proof: The assertion is local on Y' so we can take Y' to be affine and of finite type over Y. Hence Y' can be considered as a closed subspace of affine n-space over Y for some n. By Proposition 1.4, affine n-space over Y is a geometric quotient of affine n-space over X, under the induced action of G. Hence we are reduced to the case that $Y' \rightarrow Y$ is a closed immersion.

The proposition is local on Y so we can take Y affine, in which case π is open (SGA 60-61V). Hence the restriction of π to a closed subspace in its domain is an open mapping. $\blacksquare$

We will later use this proposition in the precise form of the following corollary:

<u>Corollary 1.7</u>: Let X be a G-space and $\pi: X \to Y$ a geometric quotient. Let $Y' \to Y$ be a finite type morphism and U a subspace of Y'. Then, defining X' and U' to make the following cartesian:

$$\begin{array}{ccccc} U' & \longrightarrow & X' & \longrightarrow & X \\ \downarrow & & \downarrow & & \downarrow \pi \\ U & \longrightarrow & Y' & \longrightarrow & Y \end{array}$$

the natural map

$$(\text{Closure of } U' \text{ in } X') \to (\text{Closure of } U \text{ in } Y') \underset{Y'}{\times} X'$$

is an isomorphism of the underlying topological spaces. (Closure here is taken in the sense of II.5.7)

<u>Proof</u>: The proposition is local on Y' so we can take Y' to be affine. It is also clearly sufficient to assume that U is algebraically, hence topologically, dense in Y'. Then we have

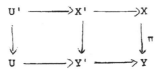

$$\begin{array}{ccc} U' & \longrightarrow & X' \\ \downarrow & & \downarrow \\ U & \longrightarrow & Y' \end{array}$$

a cartesian diagram of schemes with $U \longrightarrow Y'$ a topologically dense open immersion and $X' \longrightarrow Y'$ an open mapping (by 1.6). It is then clear that U' is topologically dense in X'. ∎

1.8: In this section, we have proved only the bare minimum necessary for the proof of the Chow lemma.

In general, Deligne has shown that in the category of separated algebraic spaces, quotients under finite group actions always exist - that is, quotient maps $X \longrightarrow X/G$ which are both geometric and categorical quotients (although not always universal geometric or categorical quotients).

We briefly indicate the proof: Let X and Y be algebraic spaces on which a finite group G acts. Given any G-map $f:X \longrightarrow Y$, if we write Y^σ and X^σ for the fixed-point locus of any σ in G, $X^\sigma \subset f^{-1}(Y^\sigma)$. We say f is fixed-point-reflecting (FPR) if this is an equality for each $\sigma \in G$. The existence of geometric quotients depends on two lemmas: 1) Given a separated algebraic space X, on which a finite group G acts, there is a scheme U, a disjoint union of affine schemes, an action of G on U, and an etale surjective G-map $U \longrightarrow X$ which is FPR. (The proof of this uses Theorem II.6.4). 2) If $f:X \longrightarrow Y$ is an etale FPR G-map of separated G-spaces and geometric quotients X/G, Y/G and f/G exist, then f/G is etale. Given these, we start with a G-space X, use 1) to find an affine covering $U \longrightarrow X$, compatible with the G-action, form $R = U \times_X U$ and note that each map $R \rightrightarrows U$ is a FPR etale covering. Since R, and U are disjoint unions of affine schemes, R/G, U/G and f/G all exist and $R/G \rightrightarrows U/G$ is an etale equivalence relation. Its quotient is X/G.

Thus the category of separated algebraic spaces is closed under finite group quotients. Inside this category, the category of quasiprojective schemes (i.e,,the category of all schemes quasiprojective over a given noetherian separated base algebraic space) is closed under finite group actions. And, of course, the category of affine schemes is so closed. But the category of separated schemes is not so closed, as the example on p.14 shows. This is another justification for the assertion that the notion of algebraic space (rather that the notion of scheme) is the "right" generalization of the notion of quasiprojective scheme.

We abstain (for the moment at least) from the task of giving a general outline of the theory of finite group actions (for reasons both of time and length). The general theory exists in print in XXVI, XXXV, SGA V, Bourbaki: Alg. Comm. Chap. 5, Serres's Groupes Algebriques, and Mumford's Abelian Varieties (Oxford 1970).

The fact that quasiprojective varieties admit quasiprojective quotients under finite group actions seems to be "generally known" but it is hard to give a specific reference, and hence the indication of proof of this above.

All the theorems and definitions above can be generalized to some extent to the case of algebraic groups. But there are indications that the problems involved in finding quotients of algebraic spaces by general algebraic groups can be reduced to problems involving finite groups. (See Seshadri XXXI.)

2. Symmetric powers of Projective Spaces

The object here is to take a finite etale map $U \to V$, of degree n, and dominate it by a particular map $U_1 \to U \to V$ which is finite etale and such that $U_1 \to V$ is a geometric quotient of U_1 under an action by the symmetric group S_n of permutations of n letters. Then for a given projective space P we defing symmetric powers $P^n \to S^n P$ of P and show every immersion $U \to P$ of the U above induces a cartesian diagram

$$\begin{array}{ccc} U_1 & \longrightarrow & P^n \\ \downarrow & & \downarrow \\ V & \longrightarrow & S^n P \end{array}$$

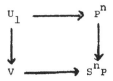

Construction 2.1: Let $\pi : U \to V$ be a finite etale map of affine algebraic spaces. Then π is affine so $U = \underset{\sim}{\mathrm{Spec}}\ \pi_* O_U$ where $\pi_* O_U$ is a coherent locally free O_V-algebra. We assume π is of degree n, i.e., that $\pi_* O_U$ is locally free of dimension n over O_V.

Remark 2.2: The simplest example here is to take U to be the disjoint sum of n copies of V (the trivial case). But in general, assertions about this situation which are local on V can always be reduced to this trivial case, since in any local argument we can replace V by the affine spectrum of a strict hensel local ring, in which case, the hensel property implies that U is a disjoint sum of n copies of V.

Consider the n-fold product $\tilde U = U \underset{V}{\times} U \underset{V}{\times} \cdots \underset{V}{\times} U$. The symmetric group S_n acts on $\tilde U$ by permutation of factors. For each $\sigma \in S_n$, the fixed point locus of the action of σ on $\tilde U$, which we denote $\tilde U_\sigma$, is defined by the cartesian diagram

$$\begin{array}{ccc} \tilde U_\sigma & \longrightarrow & \tilde U \\ \downarrow & & \downarrow{\scriptstyle 1 \times \sigma} \\ \tilde U & \xrightarrow{\Delta} & \tilde U \times \tilde U \end{array}$$

or equivalently by the cartesian diagram

$$\begin{array}{ccc} \tilde U_\sigma & \longrightarrow & \tilde U \\ \downarrow & & \downarrow{\scriptstyle 1 \times \sigma} \\ \tilde U & \xrightarrow{\Delta} & \tilde U \underset{V}{\times} \tilde U \end{array}$$

Since $\tilde{U} \to V$ is etale and separated, Δ is open and closed. Hence $\tilde{U}_\sigma$ is a component of $\tilde{U}$.

<u>Notation</u>: $U_1 = \tilde{U} - \bigcup\limits_{\sigma \in S_n} \tilde{U}_\sigma$. U_1 is then a component of $\tilde{U}$, and $U_1 \to V$ is etale, finite of degree $n!$ and surjective. In the action of S_n on $\tilde{U}$, U_1 is stable (since its complement is stable) and in fact S_n acts freely on U_1. Indeed, $U_1 \to V$ is a geometric quotient of U_1 by S_n. To see all of this, we can use remark 2.1 to reduce to the trivial case, where it is easy. Intuitively, U_1 is the set of all n-tuples $(u_1, u_2, \ldots, u_n)$, $u_i \in U$, $u_i \neq u_j$, $\pi(u_i) = \pi(u_j)$, where π denotes the original map $U \to V$.

Finally we note that there is a cartesian diagram

$$\begin{array}{ccc} \coprod\limits_n U_1 & \xrightarrow{\ \sigma\ } & U \\ \Big\downarrow{\scriptstyle \tau} & & \Big\downarrow{\scriptstyle \pi} \\ U_1 & \xrightarrow{\ \pi_1\ } & V \end{array}$$

with π, π_1 the projection maps, $\coprod\limits_n U_1$ a disjoint sum of n copies of U_1 with τ the disjoint sum of the identity maps, and σ the map which on the ith copy of U_1 is the composite:

$$U_1 \hookrightarrow \tilde{U} = U \underset{V}{\times} U \underset{V}{\times} U \ldots \underset{V}{\times} U \xrightarrow{\ i^{th} \text{ projection}\ } U$$

(Again, using 2.2, this is clear in the trivial case.)

2.3: Symmetric Powers of Projective Spaces

Definition: Let P be a quasi-projective scheme (over Spec Z). Let P^n denote the n-fold power $P \times P \times \ldots \times P$. The symmetric group S_n acts on P^n by permuting factors. P^n is also quasi-projective so by Proposition 1.5, a geometric quotient exists, which we call $\pi : P^n \to S^n P$. $S^n P$ is called the n-fold symmetric power of P, and is again quasiprojective.

For $\sigma \in S_n$, let $J_\sigma \to P^n$ be the fixed point locus of σ, and $K_\sigma = P^n - J_\sigma$ its open complement. Let $K = \bigcap_{\sigma \in S_n} K_\sigma$. Then K is an open subspace of P^n on which S_n acts freely. (K is clearly stable under the action of S_n.) Let L be the quotient of K under S_n. Then by proposition 1.5, there is a cartesian diagram

$$\begin{array}{ccc} K & \hookrightarrow & P^n \\ \downarrow & & \downarrow \\ L & \hookrightarrow & S^n P \end{array}$$

Since S_n acts freely on K, $K \to L$ is etale.

2.4 (Combining 2.1 and 2.3): Let $\pi : U \to V$ be a finite etale map of degree n of affine algebraic spaces and U_1 be as above so $\pi_1 : U_1 \to V$ is a geometric quotient of U_1 under the group S_n. S_n acts freely on U_1 so V can be identified as the quotient of the etale equivalence relation $S_n \times U_1 \rightrightarrows U_1$. (Example 1.2).

Let $U \to P$ be an immersion of U into a projective space P. Then there is an immersion $\tilde{U} \to P^n$, so an immersion $U_1 \to P^n$.

Let $K \to P^n$ be the subspace of P^n on which S_n acts freely (as defined above in part B) and L be the quotient of K under the action of S_n. S_n acts freely on K so L is the quotient of the etale equivalence relation $S_n \times K \rightrightarrows L$.

Clearly, $U_1 \longrightarrow P^n$ factors $U_1 \xrightarrow{i} K \longrightarrow P^n$ with i an immersion. Hence the diagram

$$\begin{array}{ccccc} S_n \times U_1 & \rightrightarrows & U_1 & \longrightarrow & V \\ \downarrow{\scriptstyle 1 \times i} & & \downarrow{\scriptstyle i} & & \downarrow{\scriptstyle j} \\ S_n \times K & \rightrightarrows & K & \longrightarrow & L \end{array}$$

defines a map $J : V \to L$. Since each square

$$\begin{array}{ccc} S_n \times U_1 & \longrightarrow & U_1 \\ \downarrow{\scriptstyle 1 \times i} & & \downarrow{\scriptstyle i} \\ S_n \times K & \longrightarrow & K \end{array}$$

is cartesian, and i is an immersion, the square

$$\begin{array}{ccc} U_1 & \longrightarrow & V \\ \downarrow{\scriptstyle i} & & \downarrow{\scriptstyle j} \\ K & \longrightarrow & L \end{array}$$

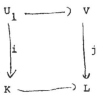

is cartesian and j is an immersion. (By descent theory for subspaces.) Combining this result with the fact that

$$\begin{array}{ccc} K & \longrightarrow & P^n \\ \downarrow & & \downarrow \\ L & \longrightarrow & S^n P \end{array}$$

is cartesian, we finally have a cartesian diagram

$$\begin{array}{ccc} U_1 & \xrightarrow{\ i\ } & P^n \\ \downarrow & & \downarrow \\ V & \xrightarrow{\ j\ } & S^n P \end{array}$$

with all the maps as identified above, and, in particular, with j an immersion.

2.5: Generalization to Algebraic Spaces Over an Arbitrary Base

Finally, not that this entire section 2 can be redone taking all spaces to be separated and of finite type over a noetherian base algebraic space S, interpreting affine to mean affine over S, quasiprojective to mean quasiprojective over S, etc.

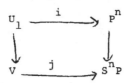

2.6: Another application of the notion of symmetric power of a quasiprojective variety is the proof of Theorem II.6.4, page 129. Recall this theorem states: Let X be an algebraic space and x $\longrightarrow$X a point of X. Then there is an affine scheme U

and an etale map $U \longrightarrow X$ such that $x \longrightarrow X$ factors $x \longrightarrow U \longrightarrow X$.

<u>Proof</u>: Given an etale covering $W' \xrightarrow{\pi'} X$ of X, let $y = \pi'^{-1}(x)$ and $W \longrightarrow W'$ an affine open subscheme of W' containing y. Then $W \xrightarrow{\pi} X$ is quasiaffine (since each map $W \times_X W \rightrightarrows W$ is, applying I.5.12) and etale and there is a commutative diagram

$$\begin{array}{ccc} y & \longrightarrow & W \\ \downarrow & & \downarrow \\ x & \longrightarrow & X \end{array}$$

with $k(y) \longleftarrow k(x)$ a separable field extension, of degree n, say. Given any finite separable extension $L/k(y)$, there is an affine etale extension W'' of W and a cartesian diagram

$$\begin{array}{ccc} \operatorname{Spec} L & \dashrightarrow & W'' \\ \downarrow & & \downarrow \\ y & \longrightarrow & W \end{array} \qquad \text{(applying SGA I).}$$

Hence we can assume $k(y)$ is also normal over $k(x)$. Let $G = \{\sigma_1, \ldots, \sigma_n\}$ be the Galois group.

Form the n-fold symmetric product of W/X, $\mathrm{Symm}^n_X(W)$, which exists since W/X is quasiaffine. For each map σ_i in G, there is a map $y \xrightarrow{\sigma_i} y \longrightarrow W$ and we let $y \longrightarrow \mathrm{Symm}^n_X(W)$ be the map given by $y \xrightarrow{\sigma} W^n$, $\sigma = \langle \sigma_1, \ldots, \sigma_n \rangle$.

y is a G-space and the map $y \longrightarrow \mathrm{Symm}^n_X(W)$ is G-invariant, so it factors through the quotient $y/G = x$:

$$\begin{array}{ccccc} y & \longrightarrow & x & \longrightarrow & \mathrm{Symm}^n_X(W) \\ & \searrow & \Big\downarrow 1 & & \Big\downarrow \pi \\ & & x & \longrightarrow & X \end{array}$$

Hence $\mathrm{Symm}^n_X(W)$ contains a point x' with $\pi(x') = x$, and $k(x') = k(x)$.

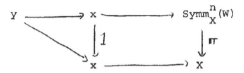

Now we claim $\text{Symm}_X^n(W) \xrightarrow{\pi} X$ is etale at x'. Since the construction is local on X, we can assume X is a strict Hensel local ring and hence W is a sum of n copies of X. Then $W \times_X W \times_X \cdots \times_X W$ is a sum of n^n copies of X, on which the group of permutations of factors acts freely. Thus the quotient $\text{Symm}_X^n(W)$ is a sum of copies of X so etale over X.

Finally, we take U to be an affine open subspace of $\text{Symm}_X^n(W)$ containing the point x'. ∎

3. Chow's Lemma

Theorem 3.1: Let $f:E \to Y$ be a separated finite type morphism of separated noetherian algebraic spaces. Then there exists an algebraic space $\overline{V}$ and a map $g:\overline{V} \to E$ such that

1) $G:\bar{V} \to E$ is projective and birational (hence surjective)

and 2) $fg:\bar{V} \to E \to Y$ is quasiprojective.

Proof: The proof of Chow's lemma first involves two easy reductions. Then there is a bit of construction and finally there are two hard facts to prove. The first of these facts has been done in Chapter II. The proof of the second is adapted from the proof of the Chow lemma in EGA II.5.6.

First Reduction: We can take E to be irreducible. By II.8.5, there is a projective, surjective, and birational map $W \to E$ with $W = \coprod E_i$ a finite disjoint union of the irreducible components of E. If for each irreducible E_i we can find an appropriate $\bar{V}_i \to E_i$ satisfying the theorem for E_i, then the map $\coprod \bar{V}_i \to E$ will satisfy the theorem for E.

Second Reduction: We can ignore Y. We do this by making the convention that from now on, affine means affine over Y, projective means projective over Y, etc. (Indeed, once we choose an affine covering $X \to E$ and a projective immersion $X \to P$, all the rest of the proof is local on Y, so we are just suppressing mention of an affine noetherian base scheme.)

Construction: Let $X \to E$ be an etale covering of E by an affine noetherian algebraic space X. By I.5.20, there is a dense open subspace V of E with $U = V \times_E X \to V$ finite and etale. E is noetherian so we can take V affine. E is irreducible so $U \to V$ has a well-defined degree n. Let $X \to P$ be an open immersion of X in a projective space P. Then the immersion $U \to X \to P$ gives an induced immersion $V \to S^nP$. (Where S^nP, a projective space, is the n-fold symmetric power of P--see Section 2.)

The immersion $V \to S^nP$ factors $V \to E \times S^nP \to S^nP$ where $V \to E \times S^nP$ is the product of the immersions, so an immersion, and the map $E \times S^nP \to S^nP$ is the second projection. Let $\bar{V}$ be the geometric closure of V in $E \times S^nP$. Then we have a string of maps

$$V \longrightarrow \bar{V} \longrightarrow E \times S^nP \longrightarrow S^nP \qquad (*)$$

We now take the composite

$$\bar{V} \longrightarrow E \times S^nP \longrightarrow E$$

to be the desired projective birational cover of E. It is projective since S^nP is projective and $\bar{V} \to E \times S^nP$ is a closed immersion.

To check the birationality, consider the cartesian squares

$$\begin{array}{ccccccc} V & \longrightarrow & (?) & \longrightarrow & V \times S^nP & \longrightarrow & V \\ \downarrow & & \downarrow & & \downarrow & & \downarrow \\ V & \longrightarrow & \bar{V} & \longrightarrow & E \times S^nP & \longrightarrow & E \end{array}$$

where, since $V \times S^nP \to E \times S^nP$ is etale surjective, (?) is the closure of V in $V \times S^nP$. But V is closed in $V \times S^nP$ since

$$\begin{array}{ccc} V & \longrightarrow & V \times S^n P \\ \downarrow & & \downarrow \\ S^n P & \xrightarrow{\Delta} & S^n P \times S^n P \end{array}$$

is cartesian and $S^n P$ is separated. Hence $V = \bar{V} \times_E V$ so $\bar{V} \to E$ is birational.

The hard part of the proof now comes in showing that $\bar{V}$ is quasi-projective. We do this by showing that the map $\bar{V} \to S^n P$ is quasiaffine.

By II.6.15, it is sufficient to show that $\bar{V} \to S^n P$ is separated, of finite presentation, and quasifinite. The map is clearly separated, of finite presentation and quasicompact, so it is sufficient to show that for any point $q \in S^n P$, the topological inverse image $|\bar{q}| = |\bar{V} \times_{S^n P} q|$ is a discrete space.

To show this, we go back to the string of maps (*) and take the pullback of everything by the geometric quotient $P^n \to S^n P$:

$$\begin{array}{ccccccc} U_1 \to \bar{U}_1 \to & \bar{V} \times_{S^n P} P^n & \longrightarrow & E \times P^n & \longrightarrow & P^n \\ \downarrow & \downarrow & & \downarrow & & \downarrow \\ V & \longrightarrow \bar{V} & \longrightarrow & E \times S^n P & \longrightarrow & S^n P \end{array}$$

where we have used 2C to identify $U_1 = V \times_{S^n P} P^n$. $\bar{U}_1$ is here the algebraic closure of the immersion $U_1 \to E \times P^n$. Since $\bar{V} \times_{S^n P} P^n \longrightarrow E \times P^n$ is a closed subspace containing U_1, it contains $\bar{U}_1$. $\bar{U}_1$ may be identical with $\bar{V} \times_{S^n P} P^n$ --this is not clear--but by 1.7 it is clear that the map $\bar{U}_1 \to \bar{V} \times_{S^n P} P^n$ is an isomorphism of the underlying topological spaces.

We wish to show that the map $\bar{V} \to S^n P$ has finite discrete fibers. Since $P^n \to S^n P$ is finite, and

$$\begin{array}{ccc} \bar{V} \times_{S^n P} P^n & \longrightarrow & P^n \\ \big\downarrow & & \big\downarrow \\ \bar{V} & \longrightarrow & S^n P \end{array}$$

is cartesian, it is sufficient to show that $\bar{V} \times_{S^n P} P^n \to P^n$ has finite discrete fibers. Since $\bar{U}_1 \to \bar{V} \times_{S^n P} P^n$ gives an isomorphism of the underlying topological spaces, it is sufficient to show that $\bar{U}_1 \to P^n$ has finite discrete fibers. Hence we are done once we have proved the following lemma:

<u>Lemma</u>: $\bar{U}_1 \to P^n$ is an immersion.

<u>Notation</u>: We have the open immersion $X \to P$. Let P^n_i be the product $P \times P \times \ldots \times X \times \ldots \times P$, with all P's, except for an X in the ith slot. Then $P^n_i \to P^n$ is an open immersion and the disjoint union $\coprod_i P^n_i \to P^n$ is a union of open subsets of P^n.

<u>Claim</u>: The image of $\bar{U}_1 \to P^n$ lies inside the image of $\coprod_i P^n_i \to P^n$.

First let's see why the truth of this claim will prove the lemma.

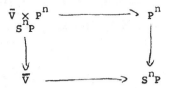

The image of $\coprod_i P^n_i \to P^n$ is open, and if the claim is true, it is sufficient to check, for each $i = 1,2,\ldots,n$, letting $\bar{U}_{1i}$ be the product $\bar{U}_1 \times_{P^n} P^n_i$, that $\bar{U}_{1i} \to P^n_i$ is a subspace. Thus we can take the string of maps

$$U_1 \to \bar{U}_1 \to E \times P^n \to P^n$$

and replace it by

$$U_1 \to \bar{U}_{1i} \to E \times P^n_i \to P^n_i$$

where $\bar{U}_{1i}$ is the closure of U_1 in $E \times P^n_i$. (Note the image of $U_1 \to P^n$ lies inside P^n_i since U_1 lies inside $X^n = \bigcap_i P^n_i$.)

Consider now the map $P^n_i \xrightarrow{\pi_1} E$ which is the ith projection followed by the covering $X \to E$. Since E is separated, the graph $P^n_i \longrightarrow E \times P^n_i$ is a closed immersion. The subspace $U_1 \to E \times P^n_i$ lies inside this closed subspace. I.e., there is a commutative diagram:

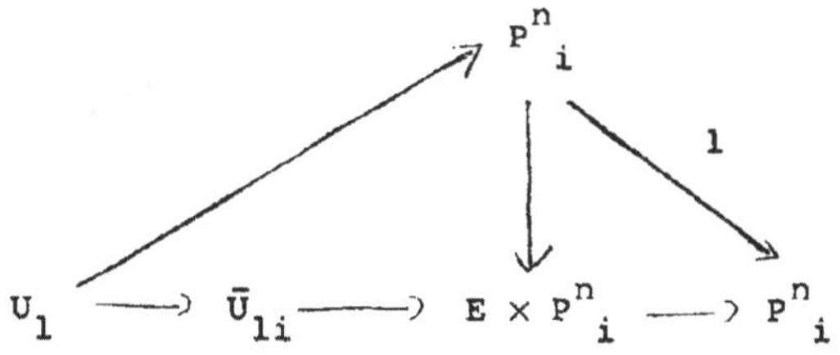

Hence the closure $\bar{U}_{1i}$ of U_1 in $E \times P^n_i$ factors $\bar{U}_{1i} \to P^n_i \to E \times P^n_i$

so the composite $\bar{U}_{1i} \to P^n_i$ is a subspace.

Hence, retracing our steps, $\bar{U}_1 \to P^n$ is a subspace so $\bar{V} \to S^n P$ is quasiaffine.

<u>Proof of Claim</u>: We have a diagram

$$\underset{i}{\bigsqcup} P^n_i$$

$$U_1 \longrightarrow \bar{U}_1 \longrightarrow E \times P^n \xrightarrow{\;\;\pi_{_{I}}\;\;} P^n$$

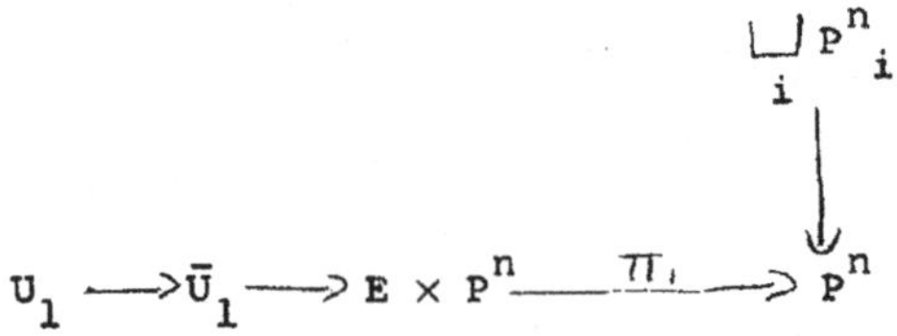

Consider also the diagram

$$U_1 \longrightarrow \bar{U}_1 \longrightarrow E \times P^n \xrightarrow{\;\;\pi_{_{2}}\;\;} E \uparrow^{\pi} X$$

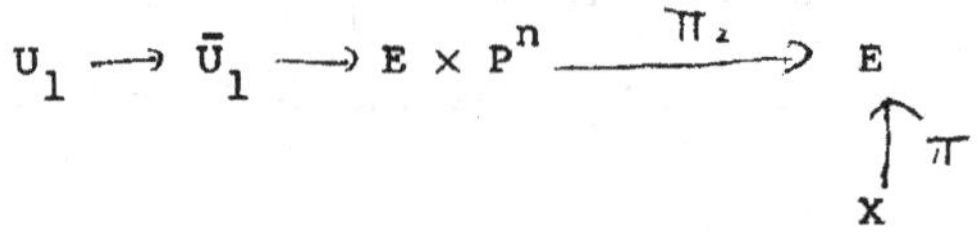

We now combine these two and form lots of cartesian squares:

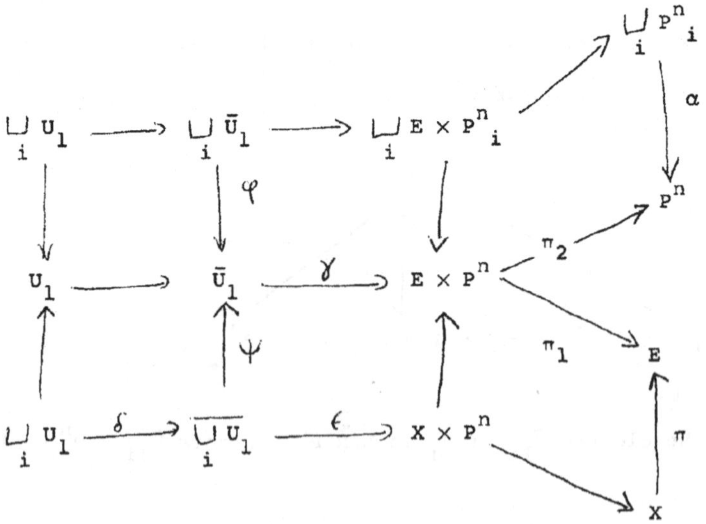

In identifying the various pullbacks we have used the notation $Y \to \bar{Y} \to Z$ to denote the closure of an immersion $Y \to Z$. Since the right hand maps are etale, closure commutes with pullback. Finally, the identification of the lower left hand corner as $\bigsqcup_i U_1$, the disjoint sum of n copies of U_1 is from section 2. Several maps are labeled to facilitate the following discussion.

The claim is that the image of $\pi_2\gamma:\bar{U}_1 \to P^n$ lies inside the image of $\alpha: \bigsqcup_i P^n_i \to P^n$. α is a covering of some subspace of P^n, by open subspaces, so φ is an open covering of some part of $\bar{U}_1$. Hence, to prove the claim we need only show that φ is surjective.

Now $\pi:X \to E$ is an etale covering, so ψ is an etale covering, hence surjective. To prove that φ is surjective, it would be sufficient to find a map $\beta: \overline{\bigsqcup_i U_1} \to \bigsqcup_i \bar{U}_1$ such that $\varphi\beta = \psi$. Unfortunately, such a β seems impossible to find. But with a slight modification this idea works.

The map $\epsilon\delta: \bigsqcup_i U_1 \to X \times P^n$ is an immersion and on the ith summand $\epsilon\delta_i:U_1 \to X \times P^n$ is the product of $U_1 \hookrightarrow U \times U \times \ldots \times U \xrightarrow{i^{th} \text{ projection}} X$ and the immersion $U_1 \to P^n$. Let $\bar{\bar{U}}_1 \to X \times P^n$ denote the closure of this i^{th} immersion $\epsilon\delta_i$. Then there is a commutative diagram

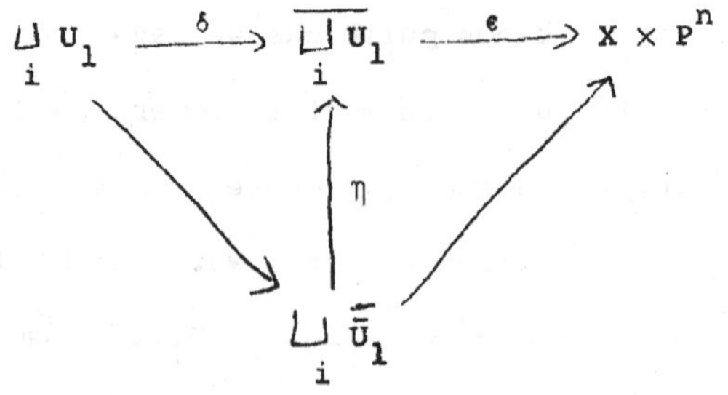

where η is surjective.

Now let's rewrite out big diagram inserting the map η:

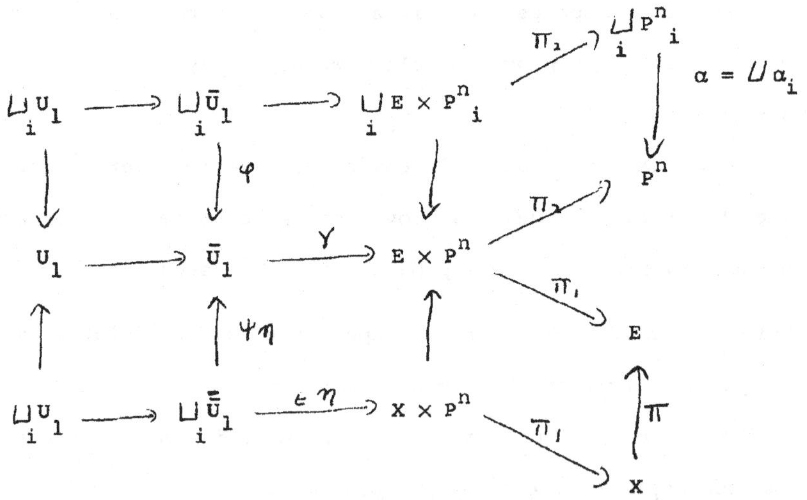

Now we can use our idea--to show φ is surjective we show that $\psi\eta$ (which is surjective) factors through φ. We do this by showing that, for each $i = 1,2,\ldots,n$, the ith copy of $\bar{\bar{U}}_1$ in $\bigsqcup_i \bar{\bar{U}}_i$ maps to the ith copy of $\bar{U}_1$ in $\bigsqcup_i \bar{U}_1$. Fix i then, and write $\bar{U}_{1i}$ and $\bar{\bar{U}}_{1i}$ for these i^{th} copies.

Since $\bar{U}_{1i} = (\bar{U}_1) \times_{P^n} (P^n{}_i)$, to give a map $\bar{\bar{U}}_{1i} \to U_{1i}$, it is sufficient to take the given map $\psi\eta_i : \bar{\bar{U}}_{1i} \to \bar{U}_1$ and to find a new map $\lambda : \bar{\bar{U}}_{1i} \to P^n{}_i$ such that

$$\begin{array}{ccc} \bar{\bar{U}}_{1i} & \xrightarrow{\lambda} & P^n{}_i \\ {\scriptstyle \psi\eta_i}\downarrow & & \downarrow{\scriptstyle \alpha_i} \\ \bar{U}_1 & \xrightarrow{\pi_2\gamma} & P^n \end{array}$$

commutes. Since α_i is an open immersion, what we need to do is to show that the image of $\pi_2\gamma\psi\eta_i : \bar{\bar{U}}_{1i} \to P^n$ lies inside $P^n{}_i$. To see this we will prove that the following diagram commutes

$$\begin{array}{ccc} \bar{\bar{U}}_{1i} & \xrightarrow{\pi_2\gamma\psi\eta_i} & P^n \\ {\scriptstyle \pi_1\epsilon\eta_i}\downarrow & & \downarrow{\scriptstyle i^{th}\ \text{projection}} \\ X & \xrightarrow{\text{injection}} & P \end{array} \qquad (**)$$

Note now that the injection $U_1 \to \bar{\bar{U}}_{1i}$ is a categorical epimorphism in the category of separated algebraic spaces (II.5.8). and P is separated. Thus to show (**) commutes, it is sufficient to show

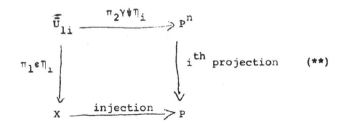

$$\begin{array}{ccc} U_1 & \xrightarrow{\text{injection}} & P^n \\ {\scriptstyle i^{th}\ \text{projection}}\downarrow & & \downarrow{\scriptstyle i^{th}\ \text{projection}} \\ X & \xrightarrow{\text{injection}} & P \end{array}$$

is a commutative diagram, which is clear. Hence the claim, hence the theorem. ∎

4. The Finiteness Theorem

Theorem 4.1: (The Finiteness Theorem. Cf. [EGA 3.2.1]) Lef $f:X \to Y$ be a proper morphism of noetherian separated algebraic spaces and F a coherent sheaf on X. Then $R^q f_*(F)$, $q \geq 0$, are coherent O_Y-modules.

Proof: Clearly the set of all coherent F, with $R^q f_*(F)$ coherent for all $q \geq 0$, is exact in the sense of devissage. Hence it is sufficient to show, for every integral $W \to X$, that there is a coherent sheaf F on X, with W = Supp F, and $R^q f_*(F)$ coherent for all $q \geq 0$. Thus we can assume X is integral, and the problem is to find a coherent sheaf F on X, with Supp F = X, and $R^q f_*(F)$ coherent for all $q \geq 0$.

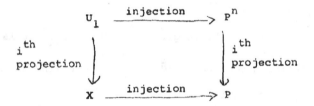

By Chow's Lemma, there is an algebraic space $X' \xrightarrow{\pi} X$ with π surjective, projective and birational and with the composition $h:X' \to X \to Y$ projective. Since X is integral, we can assume X' is integral. Let $F = \pi_* \mathcal{O}_{X'}(n)$ for some very large n (where $\mathcal{O}_{X'}(1)$ is an ample sheaf relative to $\pi: X' \to X$).

Since π is projective, F is coherent. Also $\pi^*\pi_*(\mathcal{O}_{X'}(n)) \to \mathcal{O}_{X'}(n)$ is surjective since n is large. Since the generic point of X' is mapped onto the generic point $x \in X$, $F_x \neq 0$. Also $h: X' \to Y$ is projective so $R^q h_*(\mathcal{O}_{X'}(n))$ is coherent for $q \geq 0$. (We are here applying II.7.11.)

There is a spectral sequence

$$E_2^{pq} = R^p f_*(R^q \pi_*(\mathcal{O}_{X'}(n))) \Rightarrow R^{p+q} h_*(\mathcal{O}_X(n))$$

Since n is large, $R^q \pi_*(\mathcal{O}_{X'}(n)) = 0$ for $q > 0$ so

$$R^p f_*(\pi_*(\mathcal{O}_{X'}(n))) = R^p h_*(\mathcal{O}_{X'}(n))$$

or $R^p f_*(F) = R^p h_*(\mathcal{O}_{X'}(n))$. Hence $R^p f_*(F)$ are coherent for all $p \geq 0$. ∎

CHAPTER FIVE

FORMAL ALGEBRAIC SPACES

1. Affine Formal Schemes 204

2. Formal Algebraic Spaces 215

3. The Theorem of Holomorphic Functions 224

4. Applications to Proper Morphisms 233

5. Completions of Modules of Homomorphisms 241

6. The Grothendieck Existence Theorem 245

1. Affine Formal Schemes

Throughout this chapter, all rings will be assumed to be Noetherian.

Definition 1.1: A topological ring R is called adic if R has an ideal I, wuch that the topology on R is the I-adic topology, and R is separated and complete in this topology. (I.e., $R = \varprojlim_{n} R/_{I^n}$, where each $R/_{I^n}$ has the discrete topology. In particular this means that each ideal I^n is both open and closed).

Any such ideal I is called an ideal of definition of R. If I_1 and I_2 are two ideals of definition of R, there must be integers

n_1 and n_2 with $I_1^{n_1} \subseteq I_2$ and $I_2^{n_2} \subseteq I_1$. We will sometimes write (R,I) to specify a particular ideal of definition. An adic ring is trivially adic (of discrete) if I is nilpotent.

Since R is a noetherian adic ring, R has a unique largest ideal of definition I. (For any ideal of definition J of R, I is the radical of J.) We define the nth truncation of R to be the discrete ring R/I^n, where I is the unique largest ideal of definition of R.

A map of adic rings is a continuous ring homomorphism. (Note $f:(R,I) \to (S,J)$ is continuous iff there is an integer n with $f^{-1}(J) \supseteq I^n$.)

Let (R,I) be an adic ring and M an R-module. M is a continuous R-module if $M = \varprojlim_n M/I^nM$.

Proposition 1.2: (Krull) Let (R,I) be an adic ring (noetherian as always) and M an R-module of finite type. Then M is continuous.

Proof: See EGA 0_I.7.3.3 ∎

A map $f:R \to S$ of adic rings is an adic map if for some (hence for every) ideal of definition I of R, the S-ideal $f(I)\cdot S$ is an ideal of definition of S.

Proposition 1.3: Let $f:R \to S$ be an adic map. Then S is a continuous R-module. Let M be any S-module. Then M can be considered as an R-module (by $r\cdot m = f(r)\cdot m$, $r \in R$, $m \in M$) and if M is continuous as an S-module, M is also continuous as an R-module. ∎

Let (R,I) be an adic ring and M an R-module. The completion $\hat{M}$ of M is defined by $\hat{M} = \varprojlim_n M/I^nM$.

Proposition 1.4: With the above notation:

(1) $\hat{M}$ is a continuous R-module

(2) $M = \hat{M}$ iff M is a continuous R-module.

(3) $\wedge$ is a right exact functor from R-modules to continuous R-modules, left adjoint to the inclusion functor (continuous R-modules) $\hookrightarrow$ (R-modules). $\wedge$ is left exact on modules of finite type.

Proof: (1) and (2) are simple consequences of the definitions. The last assertion is specifically: for all R-modules N and all continuous R-modules M

$$\mathrm{Hom}_R(N,M) = \mathrm{Hom}_R(\hat{N},M) \quad .$$

This natural equivalence is clear. Hence the right exactness of $\wedge$. By EGA 0_I.7.3.3, $\wedge$ is left exact on modules of finite type. ∎

Definition 1.5: Let (R,I) be an adic ring and M and N R-modules. The complete tensor product of M and N, denoted $M \hat{\otimes}_R N$ is the completion of the usual tensor product: $M \hat{\otimes}_R N = \widehat{M \otimes_R N}$.

The complete module of homomorphisms is the completion of the usual hom-set: $\mathrm{Hom}^{\wedge}_R(M,N) = \widehat{\mathrm{Hom}_R(M,N)}$.

Of course, if M and N are R-modules of finite type, so are $M \otimes_R N$ and $\mathrm{Hom}_R(M,N)$ so these modules are already complete.

Definition 1.6: Let $f:R \to S$ be an adic map. We say f is of *finite type* if for some (hence for every) ideal of definition I of R, the associated map $R/_I \to S/_{IS}$ identifies $S/_{IS}$ as an algebra of finite type over $R/_I$.

Proposition 1.7: Let $f:R \to S$ and $g:R \to T$ be adic maps. Suppose f is of finite type. Then $S \mathbin{\hat{\otimes}_R} T$ is a noetherian adic ring and the map $T \to S \mathbin{\hat{\otimes}_R} /T$ is adic of finite type.

Proof: This is a corollary of EGA 0_I.7.5.5. ∎

Definition 1.8: Let $F:R \to S$ be a map of adic rings. f is *flat* (resp. *faithfully flat*) if the functor

$$\otimes_R S:(R\text{-modules}) \to (S\text{-modules})$$

is exact (resp. exact and faithful).

As usual, to check that f is flat it is sufficient to look at the restriction

$$\otimes_R S:(R\text{-modules of finite type}) \to (S\text{-modules of finite type}).$$

Thus it is sufficient to show that

$$\hat{\otimes}_R S:(\text{continuous } R\text{-modules}) \to (\text{continuous } S\text{-modules})$$

is exact (or exact and faithful).

Proposition 1.9: If $R \to S$ is an adic faithfully flat map of adic rings, then for every R-module M of finite type, the sequence

$$M \to S \otimes_R M \rightrightarrows S \otimes_R S \otimes_R M$$

is exact and we have the usual descent theory for such modules. (cf. I.3.2)

Proof: SGA VIII. ∎

__Definition 1.10:__ An adic map $f:R \to S$ of adic rings is __formally etale__ (resp. a __formal etale covering__) if for every ideal of definition I of R, the induced map $R/_I \to S/_{IS}$ is etale (resp. etale surjective) in the usual sense.

__Proposition 1.11:__

(1) An adic map $f:R \to S$ of adic rings is formally etale if and only if there exists an ideal of definition I of R such that $R/_I \to S/_{IS}$ is etale.

(2) For any ideal of definition I of an adic ring R, there is a one-one correspondence between formally etale maps $R \to S$ of adic rings, and etale maps $R/_I \to T$ of (discrete) rings given by $T = S/_{IS}$.

(3) If $f:R \to S$ is formally etale and I is the maximal ideal of definition of R. then $f(I)\cdot S$ is the maximal ideal of definition of S.

__Proof:__ (1) and (2) are corollaries of SGA I.8.3. (3) is a corollary of I.4.9(2). ∎

__Proposition 1.12:__

(1) Let $f:R \to S$ be a formally etale map of adic rings. Then f is flat. f is formally etale surjective iff f is also faithfully flat.

(2) Let $f:R \to S$ and $g:R \to T$ be adic maps of adic rings. Let $h:R \to S \times T$ be the induced map of R into the cartesian

product of S and T. Then h is adic, and h is formally etale iff both f and g are.

(3) A formally etale surjective map of adic rings is a universally effective epimorphism in the category of adic rings.

(4) Let

$$\begin{array}{ccc} R & \xleftarrow{f} & S \\ & \nwarrow h \quad g \nearrow & \\ & T & \end{array}$$

be a commutative diagram of adic rings with h formally etale. If if is formally etale surjective, then g is formally etale. If g is formally etale, so is f.

<u>Proof:</u>

(1) For an R-module M of finite type, $M \otimes_R S$ is complete so

$$M \otimes_R S = \varprojlim_n \left((M \otimes_R R/_{I^n}) \otimes_{R/_{I^n}} S/_{I^n S} \right) = \varprojlim_n \left(M \otimes_{R/_{I^n}} S/_{I^n S} \right).$$

Completion and $\otimes_{R/_{I^n}} S/_{I^n S}$ are both exact.

(2) This is the axiom S_1 of I.1.19 and is clear from the truth of S_1 for etale maps (I.4.5).

(3) This is analogous to axiom S_2 of I.1.19 and follows from the proof of the fact for etale maps (I.4.5) and the adjointness relation 1.4 above.

(4) This is axiom S_3 of I.1.19, again following from the similar statement for etale maps I.4.5. ∎

<u>Definition 1.13:</u> Given any ring R and ideal I of R, we can form

the <u>completion of R along I</u>, $\hat{R}_I = \varprojlim_n R/I^nR$, which is then an adic ring with ideal of definition I. If R is noetherian, so is R_I. (EGA 0_I.7.2.6)

The construction of completions has the following functorial property: there is a map $i:R \to \hat{R}_I$ of rings such that for any map of rings $j:R \to S$ where S is adic with ideal of definition J and where for some integer n, $j^{-1}(J) \supseteq I^n$, there is a unique continuous map $f:\hat{R}_I \to S$ with $fi = j$. If $f:R \to S$ is any map of rings, I an ideal of R and $J = f(I)\cdot S$, then there is a canonical induced map of completions which we denote $\hat{f}:\hat{R}_I \to \hat{S}_J$.

<u>Proposition 1.14:</u> Let $f:R \to S$ be an etale map of noetherian rings and I an ideal of R. Let $J = f(I)\cdot S$. Then the induced map $\hat{R}_I \to \hat{S}_j$ is formally etale. Also $\hat{S}_J = S \mathbin{\hat{\otimes}_R} \hat{R}_I$.

Suppose further that $f:R \to S$ is etale and faithfully flat. Then by I.3.1a, there is an exact sequence $R \to S \rightrightarrows S \otimes_R S$. But now we also have an exact sequence $R_I \to S_J \rightrightarrows S_J \mathbin{\hat{\otimes}_{R_I}} S_J$. (I.e., "completion along an ideal is local in the etale topology".)

<u>Proof:</u> The first assertion is clear. The second follows from the fact that $R_I \to S_J$ is formally etale surjective plus 1.12(3). ∎

Let R be a noetherian ring and M an R-module. Let I be an ideal of R. M then induces an R_I-module, $\hat{M} \to \varprojlim_n M/I^nM$

<u>Proposition 1.15:</u> With the notation above, if M is an R-module of finite type the canonical map $M \otimes_R \hat{R}_I \to \hat{M}$ is an isomorphism. Furthermore, $\hat{R}_I$ is a flat R-module.

Proof: EGA 0_I.7.3.3 ∎

Definition 1.16: The category of affine formal schemes is the dual of the category of adic rings. Given an adic ring R, we write Spf R for the associated affine formal scheme. If $\mathfrak{X}$ = Spf R is an affine formal scheme and R_n is the nth truncation of R, we write $\mathfrak{X}_n$ = Spf R_n, the nth truncation of $\mathfrak{X}$. In the category of affine formal schemes, $\mathfrak{X} = \varinjlim_n \mathfrak{X}_n$. Given an affine formal scheme $\mathfrak{X}$, the first truncation $\mathfrak{X}_1$ will be called the associated closed subscheme (or carrier) of $\mathfrak{X}$.

Since the categories of adic rings and affine formal schemes are dual, we can say a map of affine formal schemes is etale, or of finite type, or flat, if the associated map of rings is such. From now on, definitions will be extended back and forth between these two categories in this way without further explicit mention.

An adic map of affine formal schemes is an imbedding (resp. a closed imbedding, resp. an open imbedding) if the associated map of their carriers is an imbedding (resp. closed imbedding, resp. an open imbedding).

The Global etale topology on the category of affine formal schemes is the Grothendieck topology associated with the subcategory of formal etale maps. ("Associated" in the sense of I.1.16)

The Local etale topology on a given affine formal scheme $\mathfrak{X}$ is the restriction of the global etale topology to affine formal schemes $\mathfrak{Y}$ formally etale over $\mathfrak{X}$.

A sheaf (of sets, say) on an affine formal scheme $\mathfrak{X}$ is the usual-- a contravariant functor F:(affine formal schemes formally etale over $\mathfrak{X}$) $\to$ (Sets) satisfying the sheaf axiom. For global sheaves, we make an extra restriction: A sheaf (of sets, say) is a contravariant functor F:(Affine formal schemes) $\to$ (Sets) satisfying the sheaf axiom and also satisfying an asiom of continuity: For every affine formal scheme $\mathfrak{X}$, $F(\mathfrak{X}) = \varprojlim_n F(\mathfrak{X}_n)$. A map of sheaves in either case is a natural transformation of functors.

Proposition 1.17:

(1) Every representable functor on the category of affine formal schemes is a global sheaf.

(2) Let $\mathfrak{X}$ be an affine formal scheme and $\mathfrak{X}^{\bullet}$ the associated sheaf. Let $\mathfrak{X}^{\bullet}_n$ be the sheaves associated to the nth truncations of $\mathfrak{X}$. Then $\mathfrak{X}^{\bullet} = \varinjlim_n (\mathfrak{X}^{\bullet}_n)$ - the direct limit in the category of sheaves of sets.

Proof:

(1) The sheaf axiom is 1.12(3). Continuity is immediate from the definitions.

(2) is precisely the consequence of the continuity requirement on global sheaves. Note that the second assertion does not imply that, for an affine formal scheme $\mathfrak{Y}$, $\mathfrak{X}^{\bullet}(\mathfrak{Y}) = \varinjlim_n \mathfrak{X}^{\bullet}_n(\mathfrak{Y})$. Indeed take $\mathfrak{Y} = \mathfrak{X}$. In fact, to assert this would be to assert that the set of sections of a direct limit of

sheaves is the same as the direct limit of the sets of sections. In general this is not true. ∎

1.18: Let $\mathfrak{X} =$ Spf R be an affine formal scheme. The natural map $i: \mathfrak{X}_1 \to \mathfrak{X}$ induces (by 1.11(2)) an isomorphism of the local etale topology on $\mathfrak{X}$ and the etale topology on $\mathfrak{X}_1$. Thus the functor i^*: (Abelian sheaves on $\mathfrak{X}$) $\to$ (Abelian sheaves on $\mathfrak{X}_1$) is an equivalence of categories.

Note that the local topology on each affine formal scheme is noetherian in the sense of GT.

The structure sheaf $\mathcal{O}$ of adic rings on the category of affine formal schemes is the functor $\Gamma(\mathfrak{X}, \mathcal{O}) = R$ if $\mathfrak{X} =$ Spf R. This is a sheaf by 1.14. We write $\mathcal{O}_{\mathfrak{X}}$ for the restriction of this sheaf to a particular affine formal scheme $\mathfrak{X}$. Applying 1.11(3), there is a global sheaf of ideals $\mathcal{I} \subset \mathcal{O}$, assigning to each $\mathfrak{X} =$ Spf R, the maximal ideal of definition of R. For an affine formal scheme $\mathfrak{X}$, we have the obvious definitions of coherent and quasicoherent $\mathcal{O}_{\mathfrak{X}}$-modules. A quasicoherent $\mathcal{O}_{\mathfrak{X}}$-module F is continuous if for each $\mathfrak{Y} \to \mathfrak{X}$ formally etale, $\mathfrak{Y} =$ Spf S, $\Gamma(\mathfrak{Y}, F)$ is a continuous S-module.

Proposition 1.19: Let $\mathfrak{X}$ be an affine formal scheme. The category of quasicoherent $\mathcal{O}_{\mathfrak{X}}$-modules is an abelian category with enough injectives. If F is a quasicoherent $\mathcal{O}_{\mathfrak{X}}$-module, we have $H^n(\mathfrak{X}, F) = 0$, $n \geq 0$, where $H^n(\mathfrak{X}, F)$ is the cohomology of F as abelian sheaf in the local etale topology.

Proof: Exactly as in the case of affine schemes (I.4.15). Note we do not require that the injectives be continuous modules--partly because we do not need this extra requirement but mainly because we do not know whether enough such injectives can be found. ∎

2. Formal Algebraic Spaces

<u>Definition 2.1:</u> A (separated noetherian) <u>formal algebraic space</u> is a contravariant functor F:(Affine formal schemes) → (sets) such that

(1) F is a sheaf in the global formal etale topology

(2) There is an affine formal scheme $\mathcal{Y}$ and a map of sheaves $\mathcal{Y}^{\cdot} \to F$ such that for any global sheaf G, represented by an affine formal scheme, and map G → F, the product $G \times_F \mathcal{Y}$ of global sheaves is represented by an affine formal scheme, and the map $G \times_F \mathcal{Y}^{\cdot} \to G$ is represented by a formal etale covering.

(3) Given $\mathcal{Y}^{\cdot} \to F$ as above, the map of affine schemes $\mathcal{Y}_1 \times_F \mathcal{Y}_1 \to \mathcal{Y}_1 \times \mathcal{Y}_1$ is a closed immersion.

A morphism of formal algebraic spaces $f: \mathcal{X}_1 \to \mathcal{X}_2$ is a map of sheaves.

<u>2.2:</u> At this point, we have the analogs of a number of theorems proved in Chapter II for algebraic spaces. Since we have only defined "noetherian" "separated" formal algebraic spaces, all maps of formal algebraic spaces are "separated" and "quasicompact" the verifications of these analogs are somewhat easier than in the previous case where we tried to be more general.

In particular, the formal etale topology on the category of affine formal schemes extends to a formal etale topology on the category of formal algebraic spaces and we have the usual definitions and sorites on quasicoherent sheaves. The category of formal algebraic spaces is closed under the formation of quotients of separated formal etale equivalence relations (where $\mathcal{R} \rightrightarrows \mathcal{X}$ is separated if $\mathcal{R}_1 \to \mathcal{X}_1 \times \mathcal{X}_1$ is a closed immersion) and under the formation of fiber products $\mathcal{X} \times_{\mathcal{Z}} \mathcal{Y}$ whenever one of the maps $\mathcal{X} \to \mathcal{Z}$ or $\mathcal{Y} \to \mathcal{Z}$ is of finite type. A separated noetherian formal scheme (in the sense of EGA I.10) "is" clearly a formal algebraic space.

The notions of etale, flat and finite type morphism and immersions generalize in the obvious way to formal algebraic spaces. The notion of an affine map generalizes to the notion of formal affine map in the obvious way and if $\mathcal{X}$ is a formal algebraic space, and $\mathcal{A}$ a continuous quasicoherent $\mathcal{O}_{\mathcal{X}}$-algebra of finite type, we write $\mathcal{Y} = \underline{\mathrm{Spf}}\,\mathcal{A}$ for the associated formal algebraic space formally affine ever $\mathcal{X}$.

The construction of the completion of a noetherian affine scheme along a closed subspace extends to a local construction in formal algebraic spaces, the completion $\hat{\mathcal{X}}$ (or $\hat{\mathcal{X}}_C$) of a noetherian separated algebraic space $\mathcal{X}$ along a closed subspace C.

Proposition 2.3: Given a noetherian separated algebraic space X and a closed subspace $C = \underline{\mathrm{Spec}}\, \mathcal{O}_{X/I}$, we can consider X as a trivially adic formal algebraic space. Let $\hat{X}$ be the completion of X along C. The maps $\underline{\mathrm{Spec}}\, \mathcal{O}_{X/I^n} \to X$ induce a map $i: \hat{X} \to X$ which is adic if and only if I is nilpotent. This map has the following properties:

(1) If $Y \to X$ is an etale map and $\hat{Y}$ denotes the completion of Y along the closed subspace $Y \times_X C$, then the map $\hat{Y} \to \hat{X}$ is formally etale and the map $i_Y : \hat{Y} \to Y$ is the fiber product of the map i (in the category of formal algebraic spaces).

(2) $i_* : QCS_{\hat{X}} \to QCS_X$ is exact and faithful (where as usual QCS_T is the category of quasicoherent sheaves on T).

(3) $i^* i_* : QCS_{\hat{X}} \to QCS_{\hat{X}}$ is naturally equivalent to the identity functor. i^* is an exact functor on the subcategory of coherent $\mathcal{O}_X$-modules

Proof: (1) follows from the truth of the statement in the affine case, which is 1.14. Hence (2) and (3) are local in the formal etale topology and we can assume X is affine, in which case both are clear from 1.15 and 1.4. ∎

Definition 2.4: The canonical sheaf of ideals of definition I on the category of affine formal schemes extends to the sheaf of ideals of definition, again denoted I, on the category of formal algebraic spaces. The restriction, $I_{\mathfrak{X}}$, of I to the local formal etale topology on a fixed formal algebraic space $\mathfrak{X}$, is the sheaf of ideals of definition of $\mathfrak{X}$. The nth truncation of $\mathfrak{X}$ is defined as the trivially adic formal algebraic space $\mathrm{Spf}\ \mathcal{O}_{\mathfrak{X}}/I^n$. Equivalently, if $\mathfrak{X}$ is defined by the formal etale equivalence relation $\mathcal{R} \rightrightarrows \mathcal{U} \to \mathfrak{X}$, then $\mathfrak{X}_n$ is the quotient of the etale equivalence relation

$\mathcal{R}_n \rightrightarrows \mathcal{U}_n \to \mathcal{X}_n$. Hence $\mathcal{X} = \varinjlim_n \mathcal{X}_n$.

<u>Theorem 2.5:</u> Let $\mathcal{X}$ be a formal algebraic space. Then $\mathcal{X}$ is an affine formal scheme if and only if the first truncation $\mathcal{X}_1$ is affine. $\mathcal{X}$ is a formal scheme if and only if $\mathcal{X}_1$ is a scheme.

<u>Proof:</u> If $\mathcal{X}$ is affine, or a scheme, clearly so is $\mathcal{X}_1$. Conversely, by III.3.3 and III.3.6, if $\mathcal{X}_1$ is affine, or a scheme, so are each of the $\mathcal{X}_n$, andhence by EGA I.10.6, so is $\mathcal{X}$. ∎

<u>Definition 2.6:</u> A map $f: \mathcal{Y} \to \mathcal{X}$ of formal algebraic spaces is <u>proper</u> if f is of finite type and the associated map of carriers $f_1: \mathcal{Y}_1 \to \mathcal{X}_1$ is a proper map of algebraic spaces.

<u>Proposition 2.7:</u> Let $f: X \to Y$ be a map of noetherian separated algebraic spaces which is of finite type. Let $C \to Y$ be a closed subspace and $D = C \times_Y X = f^{-1}(C)$. Suppose the associated map $D \to C$ is proper. Then $f: \hat{X}_D \to \hat{Y}_C$ is proper. ∎

The following material on graded rings and modules and Mittag-Leffler condition is given here in preparationfor the Holomorphic Functions Theorem to be proved in the next section. We give just the minimal necessary definitions and facts for our purposes. For more details, such as proofs, the reader may consult Bourbaki Alg. Comm., Chap. III, and EGA II.2 and 0_{III}.13.

<u>Definition 2.8:</u> A <u>graded</u> <u>ring</u> R is a ring R (commutative with identity as always) given with a collection $\{R_k, k = 0,1,2,\ldots\}$ of abelian groups such that

i) R_0 is a ring

ii) Each R_i is an R_0-module

iii) For each $i,j \geq 0$, there is an R_0-homomorphism

$$a_{ij}:R_i \underset{R_0}{\otimes} R_j \to R_{i+j}$$

iv) $R = \underset{i \geq 0}{\oplus} R_i$ with the multiplication on R derived from the a_{ij} above.

Proposition 2.9: Let R be any ring and I an ideal of R. The associated graded ring, gr(R), is the graded ring $gr(R) = \underset{i}{\oplus} R_i$ defined by $R_i = R/_{I^{i+1}R}$. If R is noetherian, so is gr(R).

Proof: Bourbaki Alg Comm Chap. III, 2.10, Cor. 5.∎

Definition 2.10: Let $R = \oplus R_k$ be a graded ring. A graded R-module M is an R-module M given with a collection $\{M_k, k = 0,1,2,..\}$ of R_0-submodules of M and for each $i,j \geq 0$, an R_0-homomorphism $b_{ij}:R_i \underset{R_0}{\otimes} M_j \to M_{i+j}$, such that $M = \underset{k}{\oplus} M_k$, with the structure of M as R-module derived from the maps b_{ij}.

Proposition 2.11: Let $R = \oplus R_k$ be a graded noetherian ring and $M = \oplus M_k$ a graded R-module. Suppose M is of finite type as an R-module. Then for any integer $m \geq 0$, there exists an integer n_0 such that the map $b_{mn}:R_m \underset{R_0}{\otimes} M_n \to M_{m+n}$ is surjective for all $n \geq n_0$.

Proof: See Bourbaki Alg. Comm. III.3.1, Prop. 3. ∎

Definition 2.12: Let R be a noetherian ring and I an ideal. Let H be an R-module and $H = K_0 \supseteq K_1 \supseteq K_2 \supseteq \ldots$ be a filtration of

H by sub-R-modules. The filtration $\{K_i\}$ is called I-good if

(1) $I \cdot K_j \subseteq K_{j+1}$ for all $j \geq 0$

(2) There is an integer n_0, with $I \cdot K_n = K_{n+1}$, for all $n \geq n_0$.

Proposition 2.13: Let R be a noetherian ring, I an ideal and H an R-module of finite type. Let $\{K_i, i \geq 0\}$ be a filtration of H with $I \cdot K_i \subset K_{i+1}, i \geq 0$. Let $K = \bigoplus_i K_i$, which is a graded algebra over the ring $S = \bigoplus_i I^i$. Then K_i is an I-good filtration if and only if K is of finite type over S. In this case there is an isomorphism $\hat{H}_i \to \varprojlim H/_{I^n H} \cong \varprojlim H/_{K_i}$.

Proof: See Bourbaki Alg. Comm. III.3.1, Thm. 1 and III.3.2, Prop. 4. ∎

2.14: Let R be a noetherian ring and I an ideal. Let X be an algebraic space proper over Spec R. Let F be a coherent sheaf on X. We write $I^k F$ for the sheaf which is the kernal of the map of $\mathcal{O}_X$-modules $F \to F \otimes_R R/_I k$. $I^k F$ is a coherent $\mathcal{O}_X$-module and for each $n \geq 0$, we can form $H^n(X, I^k F)$, which is an R-module of finite type (by IV.4.1, the Finiteness Theorem). Consider the sum $E = \bigoplus_{k \geq 0} H^n(X, I^k F)$. Let $S \to \bigoplus_{k \geq 0} I^k$ where by definition $I^0 = R$. Then S is a graded ring (by $I^k \cdot I^j \to I^{k+j}$ and E is a graded S-module (by, for $i \in I^k, a \in H^n(X, I^j)$, $ia \in H^n(X, I^{k+j}F)$ is the image of a in the map of cohomology groups $H^n(X, I^j F) \to H^n(X, I^{j+k}F)$ induced by the map $I^j \to I^{j+k}$ which is multiplication by i).

Proposition 2.15: With the notation as above, S is a noetherian graded ring and E is an S-module of finite type.

Proof: Applying Bourbaki, Alg. Comm. Chap. III, 2.10, Cor. 5, S is noetherian. The second assertion is EGA III.3.3.2 whose proof there for X a scheme proper over Y carries over word-for-word to the case of algebraic spaces, given our proof of the finiteness theorem IV.4.1. ∎

Definition 2.16: Let $N = \{0,1,2,\ldots\}$ be an index set and C an abelian category with projective limits. In the following we consider projective systems $\{A_i,\varphi_{ij}\}$ of objects in C, where $\varphi_{ij}:A_j \to A_i$ for each $i < j \in N$, and $\varphi_{ij}\,\varphi_{jk} = \varphi_{ik}$, $i < j < k \in N$. Such a projective system is said to satisfy the Mittag-Leffler condition, ML, if for all $n \in N$, there is a $k_0 > n \in N$ such that $\mathrm{Image}(\varphi_{nk}(A_k)) = \mathrm{Image}(\varphi_{nk_0}(A_{k_0})) \subseteq A_n$ for all $k \geq k_0$. The simplest example of a system $\{A_i,\varphi_{ij}\}$ satisfying ML is one for which all the maps φ_{ij} are epimorphisms. In this case we say $\{A_i,\varphi_{ij}\}$ is strict. Another example is when all the objects A_n involved satisfy the descending chain condition for subobjects.

Proposition 2.17: Let $0 \to A_i \xrightarrow{u_i} B_i \xrightarrow{v_i} C_i \to 0$ be an exact sequence of projective systems of abelian groups with index set N. Suppose $\{A_i\}$ satisfies ML. Then the sequence

$0 \to \varprojlim A_i \to \varprojlim B_i \to \varprojlim C_i \to 0$ is exact.

<u>Proof:</u> EGA 0_{III}.13.2.2. ∎

<u>Proposition 2.18:</u> Let $\{K_i^\cdot\}_{i\in N}$ be a projective system of complexes of abelian groups, $K_i^\cdot = (K_i^n)_{n\in N}$, in which the derivation operator is of degree 1. For each n, there is a canonical homomorphism $h_n : H^n(\varprojlim_i K_i^\cdot) \to \varprojlim_i H^n(K_i^\cdot)$. If for each n, the system $\{K_i^n\}_{i\in N}$ satisfies ML, then each h_n is bijective.

<u>Proof:</u> This is a special case of EGA 0_{III}.13.2.3. ∎

<u>Theorem 2.19:</u> Let X be a noetherian separated algebraic space and $\{F_k\}_{k\in N}$ a projective system of coherent $\mathcal{O}_X$-modules and $F = \varprojlim F_k$. Suppose the system $\{F_k, \varphi_{hk}\}$ is strict--i.e., the maps φ_{hk} are epimorphisms in the category of $\mathcal{O}_X$-modules. Then for all $i \geq 0$, the canonical maps

$$h_i : H^i(X,F) \to \varprojlim_k H^i(X,F_k)$$

are bijective.

<u>Proof:</u> Let $Y \to X$ be an etale covering of X by an affine scheme Y. For each coherent sheaf G on X, we have the Cech complex $\Gamma(X,G) \to \Gamma(Y,G) \rightrightarrows \Gamma(Y \times_X Y,G) \overset{\to}{\rightrightarrows} \ldots$ giving a complex of $\Gamma(X,\mathcal{O}_X)$-modules $K^\cdot$ with $K^n = (Y \times_X Y \times_X \ldots \times_X Y,G)$, the product being taken n+1 times. Since each of the $Y \times_X Y \times_X \ldots \times_X Y$ are affine, the cohomology of G vanishes on them and thus the cohomology of G can be computed as the Cech cohomology of this complex. Hence $H^n(X,G) = H^n(K^\cdot)$.

Given now the projective system $\{F_i\}$ of sheaves, we have a projective system of complexes $\{K_i^{\cdot}\}$ constructed as above. The hypothesis that all of the maps $F_k \to F_h$ are epimorphisms implies that the projective systems $\{K_i^n\}_{i\in N}$ are strict, hence satisfy ML. Applying the last proposition we have canonical isomorphisms

$h_n: H^n(\varprojlim_i K_i^{\cdot}) \to \varprojlim_i H^n(X,F_k)$.

Finally it remains to show that $H^n(\varprojlim_i K_i^{\cdot}) = H^n(X,F)$. For this it is sufficient to show that the Cech complex $K^{\cdot}$ associated to F is the inverse limit of the Cech complexes $K_i^{\cdot}$ associated to the F_i's. Specifically we need to show that for each $n \geq 0$, if i_n is the map $Y \times_X Y \times_X \ldots \times_X Y \to X$ (n+1 factors), then $\varprojlim_k(i_n{}^*F_k) = i_n{}^*\varprojlim_k F_k$. But i_n is flat so $i_n{}^*$ is left exact and left exact functors commute. ∎

<u>Corollary 2.20:</u> Let X be a noetherian separated algebraic space and $C \to X$ a closed subspace. Let $\mathcal{I} \subset \mathcal{O}_X$ be the sheaf of ideals defining C. Let F be a coherent $\mathcal{O}_X$-module and write F_k for $F \otimes_{\mathcal{O}_X} \mathcal{O}_X/\mathcal{I}^k$. Let $\hat{X} = \hat{X}_C$, $i:\hat{X} \to X$ the canonical map and $\hat{F} = i^*F = \varprojlim_k F_k$. Then $H^n(\hat{X},\hat{F}) = \varprojlim_k H^n(X,F_k)$, $n \geq 0$.

<u>Proof:</u> By 2.3, $i_*\hat{F} = \varprojlim_k F_k$ and, since $i:\hat{X} \to X$ is formally affine, $H^n(\hat{X},\hat{F}) \cong H^n(X,i_*\hat{F})$. The system $\{F_k\}$ is strict and hence satisfies ML. The conclusion follows from Proposition 2.19. ∎

3. The Theorem of Holomorphic Functions

Let $f:X \to Y$ be a proper morphism of neotherian separated algebraic spaces. Let $Y' \to Y$ be a closed subspace and $X' = Y' \times_Y X$. Say $Y' = \operatorname{Spec} O_Y/_{I'}$ and $X' = \operatorname{Spec} O_X/_{J'}$ where $J' = f^*I'\cdot O_X$. We can then form $\hat{Y} = \varinjlim_n \operatorname{Spec} O_Y/_{I'^n}$ and $\hat{X} = \varinjlim_n \operatorname{Spec} O_X/_{J'^n}$ and consider the proper morphism $\hat{f}:\hat{X} \to \hat{Y}$

Let F be a coherent O_X-module, $F_k = F/_{J^kF}$ and $\hat{F} = \varprojlim_k F_k$. Consider, for each $n \geq 0$, the following $O_{\hat{Y}}$-modules.

(1) $(R^n f_*(F))^\wedge$

(2) $\varprojlim_k R^n f_*(F_k)$

(3) $R^n\hat{f}_*(\hat{F})$

There are natural maps

$$\rho_n : R^n f_*(F)^\wedge \to R^n\hat{f}_*(\hat{F})$$

$$\varphi_n : R^n f_*(F)^\wedge \to \varprojlim_k R^n\hat{f}_*(F_k)$$

$$\psi_n : R^n\hat{f}_*(\hat{F}) \to \varprojlim_k R^n f_*(F_k)$$

These are constructed as follows: Let

$$\begin{array}{ccc} X & \xrightarrow{\;j\;} & Y \\ \big\downarrow{\scriptstyle g} & & \big\downarrow{\scriptstyle f} \\ W & \xrightarrow{\;i\;} & Z \end{array}$$

be a commutative diagram of algebraic spaces with

$$i_*:(\mathcal{O}_W\text{-modules}) \to (\mathcal{O}_Z\text{-modules})$$

exact. Then there is a natural map

$$i^*(R^n f_*(F)) \overset{\alpha}{\to} R^n g_*(j^*(F)),$$

where α is the image of the canonical injection $F \to j_*j^*F$ under the transformation

$$\begin{aligned}\mathrm{Hom}_Y(F,j_*j^*F) &\to \mathrm{Hom}_Z(R^n f_*F, R^n f_*(j_*j^*F))\\ &= \mathrm{Hom}_Z(R^n f_*F, R^n(fj)_*j^*F)\\ &= \mathrm{Hom}_Z(R^n f_*F, R^n(ig)_*j^*F)\\ &= \mathrm{Hom}_Z(R^n f_*F, i_*R^n g_*(j^*F))\\ &= \mathrm{Hom}_W(i^*R^n f_*F, R^n g_*(j^*F)).\end{aligned}$$

Now for each k, n

$$\begin{array}{ccc} X_k & \hookrightarrow & X \\ \downarrow{\scriptstyle f_k} & & \downarrow{\scriptstyle f} \\ Y_k & \hookrightarrow & Y \end{array}$$

is cartesian where $Y_k = \underset{\sim}{\mathrm{Spec}}\ \mathcal{O}_Y/I^k$. Hence there are maps

$$\varphi_{n,k}:R^n f_*(F)_k \to R^n f_{k*}(F_k)$$

which by naturality are compatible for different k, giving a map

$$\varphi_n : R^n f_*(F)^\wedge \to \varprojlim_k R^n f_*(F_k)$$

Similarly the squares

$$\begin{array}{ccc} X_k & \longrightarrow & \hat{X} \\ \downarrow & & \downarrow \\ Y_k & \longrightarrow & \hat{Y} \end{array} \qquad \text{and} \qquad \begin{array}{ccc} \hat{X} & \longrightarrow & X \\ \downarrow & & \downarrow \\ \hat{Y} & \longrightarrow & Y \end{array}$$

are commutative giving the maps ρ_n and ψ_n. Consider now, for each k, the commutative squares

$$\begin{array}{ccccc} X_k & \longrightarrow & \hat{X} & \longrightarrow & X \\ \downarrow{\scriptstyle f_k} & & \downarrow{\scriptstyle \hat{f}} & & \downarrow{\scriptstyle f} \\ Y_k & \longrightarrow & \hat{Y} & \longrightarrow & Y \end{array}$$

By the naturality of the construction, the diagram

$$\begin{array}{ccc} R^n f_*(F)^\wedge & \xrightarrow{\rho_n} & R^n \hat{f}_*(\hat{F}) \\ & {\scriptstyle \varphi_{n,k}} \searrow \quad \swarrow {\scriptstyle \psi_{n,k}} & \\ & R^n f_{k*}(F_k) & \end{array}$$

(considered as $O_{\hat{Y}}$-modules) commutes for each k. Hence we have

$$\begin{array}{ccc} R^n f_*(F)^\wedge & \xrightarrow{\rho_n} & R^n\hat{f}_*(\hat{F}) \\ & \searrow \varphi_n \quad \psi_n \swarrow & \\ & \varprojlim_k R^n f_*(F_k) & \end{array}$$

commuting.

Note here that the domain and range of φ_n are both topologized as inverse limite of discrete modules.

<u>Theorem 3.1:</u> (The Holomorphic Functions Theorem) With the notation above, $R^n f_*(F)$ is a coherent $\mathcal{O}_{\hat{X}}$-module for each $n \geq 0$ and each of the maps φ_n, ρ_n, ψ_n are isomorphisms. In particular, φ_n is a topological isomorphism.

<u>Proof:</u> The entire assertion is local on Y so we can assume Y is affine. Say Y = Spec A, where A is a noetherian ring, and suppose Y' is defined by an ideal I of A. We write $F_k = F/I^{k+1}F$. The assertion can then be stated in terms of the usual cohomology A-modules:

In the diagram

$$\begin{array}{ccc} H^n(X,F)^\wedge & \xrightarrow{\rho_n} & H^n(\hat{X},\hat{F}) \\ & \searrow \varphi_n \quad \psi_n \swarrow & \\ & \varprojlim_k H^n(X_k,F_k) & \end{array}$$

we have

(1) $H^n(X,F_k) = H^n(X_k,F_k)$ $k \geq 0$ (since $X_k \to X$ is an affine map)

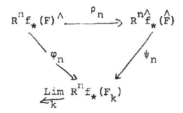

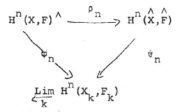

and the projective system $\{H^n(X,F_k)\}_{k \geq 0}$ satisfies ML.

(2) ψ_n is an isomorphism

(3) The kernels of the maps $H^n(X,F) \to H^n(X,F_k)$, $k \geq 0$, give an I-good filtration on $H^n(X,F)$

(4) φ_n is a topological isomorphism.

To prove this assertion, we start with the usual cohomology theory on X and consider the exact sequence:

$$H^n(X,I^{k+1}F) \to H^n(X,F) \xrightarrow{\varphi_{n,k}} H^n(X,F_k) \to H^{n+1}(X,I^{k+1}F)$$

where $\varphi_n = \varprojlim_k \varphi_{n,k}$.

To simplify notation, we write (for n fixed):

$$H = H^n(X,F) \qquad H_k = H^n(X,F_k)$$

$$R_k = \operatorname{Ker} \varphi_{n,k} \qquad Q_k = \operatorname{Image} \varphi_{n,k}$$

Thus there are exact sequences

$$0 \to R_k \to H \to H_k \to Q_k \to 0 \qquad k \geq 0$$

Let x be an element of I^m $(m \geq 0)$. The multiplication by x in I^kF is a homomorphism $I^kF \to I^{k+m}F$ and gives a homomorphism

$$\mu_{x,m} : H^n(X,I^kF) \to H^n(X,I^{k+m}F)$$

Letting S be the graded A-algebra $\bigoplus_{k\geq 0} I^k$, the multiplications define on $E = \bigoplus_{k\geq 0} H^n(X,I^kF)$ the structure of a graded module of finite type over the graded ring S which is noetherian. (Prop. 2.15)

<u>Claim:</u> The submodules R_k of H define on H an I-good filtration.

<u>Proof</u> of Claim: First of all we show that $I^m R_k \subset R_{k+m}$, the multiplication in H by an element $x \in I^m$ being the map $\mu_{x,0}$.

For any $x \in I^m$ the diagram

$$\begin{array}{ccc} I^{k+1}F & \xrightarrow{x} & I^{k+m+1}F \\ \downarrow & & \downarrow \\ F & \xrightarrow{x} & F \end{array}$$

(with the horizontal arrows being multiplication by x, and the vertical arrows the canonical injections) is commutative. Hence the corresponding diagram

$$\begin{array}{ccc} H^n(X,I^{k+1}F) & \xrightarrow{\mu_{x,m}} & H^n(X,I^{k+m+1}F) \\ \downarrow & & \downarrow \\ H^n(X,F) & \xrightarrow[\mu_{x,0}]{} & H^n(X,F) \end{array}$$

is commutative which, given the interpretation of R_k as the image of $H^n(X,I^{k+1}F) \to H^n(X,F)$ shows $I^m R_k \subset R_{k+m}$ and also shows that the graded S-module $R = \bigoplus_{k\geq 0} R_k$ is a quotient of the sub-S-module $M = \bigoplus_{k\geq 0} H^n(X,I^{k+1}F)$ of E. Thus R is an S-module of finite type which is equivalent to the claim. (Prop. 2.13) ∎ (Claim)

Consider now the graded S-module $N = \bigoplus_{k\geq 0} H^{n+1}(X,I^{k+1}F)$ defined in the same way as E above. It is an S-module of finite type and one has $Q_k \subset N_k$ for all k and as in the claim above $S_m Q_k = I^m Q_k \subset Q_{k+m}$. Thus $Q = \underset{k\geq 0}{\oplus} Q_k$ is a graded sub-S-module of N and hence is of finite type.

Let α_m be the canonical injection $I^m \to A$, which can be written $S_m \to S_o$. Since $I^{k+1}F_k = 0$, the A-module $H^n(X,F_k)$ is annihilated by I^{k+1}. Since Q_k is the image of the A-homomorphism $H^n(X,F_k) \to H^{n+1}(X,I^{k+1}F)$, Q_k, as an A-module, is also annihilated by I^{k+1}. Thus in the S-module Q, we have

$$\alpha_{k+1}(S_{k+1})\ Q_k = 0.$$

Since Q is an S-module of finite type, there are integers k_o and h such that $Q_{k+h} = S_h Q_k$ for $k \geq k_o$. This statement and the above imply that there is an integer $r > 0$ such that

$$\alpha_r(S_r)\ Q = 0$$

We now note that the canonical injection $I^{k+m}F \to I^kF$ give an A-homomorphism

$$\nu_m : H^{n+1}(X,I^{k+m}F) \to H^{n+1}(X,I^kF)$$

and for any $x \in I^m$, one has the factorization

$$\mu_{x,0} : H^{n+1}(X,I^kF) \xrightarrow{\mu_{x,m}} H^{n+1}(X,I^{k+m}F) \xrightarrow{\nu_m} H^{n+1}(X,I^kF)$$

whence we conclude that for every sub-A-module P of $H^{n+1}(X,I^kF)$, we have, in the S-module N

$$\nu_m(S_m P) = \alpha_m(S_m)P$$

<u>Claim:</u> There is an integer $m > 0$ such that $\nu_m(Q_{k+m}) = 0$ for $k \geq k_o$.

<u>Proof:</u> We can take $m \geq r$ to be a multiple of h. Since $Q_{k+m} = S_m Q_k$ for $k \geq k_o$, we have

$$\nu_m(Q_{k+m}) = \alpha_m(S_m)Q_k \subset \alpha_r(S_r)Q_k = 0 \qquad \blacksquare\text{(Claim)}$$

Consider now the commutative diagram

$$\begin{array}{ccccccc} H^n(X,F) & \to & H^n(X,F_k) & \to & H^{n+1}(X,I^{k+1}F) & \to & H^{n+1}(X,F) \\ \uparrow & & \uparrow & & \uparrow & & \uparrow \\ H^n(X,F) & \to & H^n(X,F_{k+m}) & \to & H^{n+1}(X,I^{k+m+1}F) & \to & H^{n+1}(X,F) \end{array}$$

derived from the commutative diagram

$$\begin{array}{ccccccccc} 0 & \to & I^{k+1}F & \to & F & \to & F_k & \to & 0 \\ & & \uparrow & & \uparrow & & \uparrow & & \\ 0 & \to & I^{k+m+1}F & \to & F & \to & F_{k+m} & \to & 0. \end{array}$$

From this one deduces the commutative diagram

$$\begin{array}{ccccccccccc} 0 & \to & R_k & \to & H & \to & H_k & \to & Q_k & \to & 0 \\ & & \uparrow & & \uparrow 1 & & \uparrow & & \uparrow \nu_m & & \\ 0 & \to & R_{k+m} & \to & H & \to & H_{k+m} & \to & Q_{k+m} & \to & 0 \end{array}$$

which has exact rows. As the final vertical arrow is zero for $k \geq k_o$, the image H_{k+m} in H_k is contained in $\text{Ker}(H_k \to Q_k) = \text{Im}(H \to H_k)$, and also it contains $\text{Im}(H \to H_k)$ by commutativity of the diagram. Hence they are equal. This is true for all of the images of H_k in $H_{k'}$, for $k' \geq k+m$. Hence the condition ML holds for the projective system $(H_k)_{k\geq 0}$.

We can then apply the Corollary 2.20 and the canonical map $H^n(\hat{X},\hat{F}) \to \varprojlim_k H^n(X,F_k)$ is bijective for all $n \geq 0$.

As the projective system $(H/_{R_k})_{k\geq 0}$ is strict, one can pass to the projective limit in the exact sequences

$$0 \to H/_{R_k} \to H_k \to Q_k \to 0.$$

Since $\nu_m(Q_{k+m}) = 0$, we have $\varprojlim_k Q_k = 0$ so there is a topological isomorphism $\varprojlim_k (H/_{R_k}) \xrightarrow{\sim} \varprojlim H_k$. But as the filtration (R_k) of H is I-good, it defines on H the I-adic topology. Hence $\varprojlim_k (H/_{R_k})$ is the separable completion of H for the I-adic topology. ∎

<u>Corollary 3.2:</u> Let $f:X \to Y$ be a morphism of finite type (of ordinary algebraic spaces) and F be a coherent O_X-module, whose support is proper over Y. Let $Y' \to Y$ be a closed subspace of Y, and define $\hat{Y}$, $\hat{X}$, $\hat{f}$ and F as above. Then for all $n \geq 0$ $R^n\hat{f}_*(\hat{F}) \cong (R^n f_*(F))^\wedge$.

<u>Proof:</u> We can assume that $F = u_*(G)$ where $G = u^*(F)$ is a coherent O_X-module, and Z is an appropriate closed subspace of X, whose underlying space is Supp(F) and for which $u:Z \to X$ is the canonical injection. (Proposition II.5.16) Putting $G_k = G \otimes O_{Y}/_{I^{k+1}}$ (O_Y), we have $G_k = u^*F_k$, $R^n f_*(F_k) = R^n(f\cdot u)_*(G_k)$ and $R^n f_*(F) = R^n(f\cdot u)_*(G)$ so $R^n\hat{f}_*(\hat{F}) \to R^n(f\cdot u)^\wedge_*(G)$, and now the above theorem applies. ∎

4. Applications to Proper Morphisms

Theorem 4.1: (The Connectedness Theorem of Zariski.) Let $f:X \to Y$ be a proper morphism of noetherian separated algebraic spaces. Let

$$\begin{array}{ccc} X & \xrightarrow{g} & Z \\ & f\searrow & \swarrow h \\ & Y & \end{array}$$

be its Stein factorization (II.5.3)

Then h is a finite morphism and g is a proper Stein morphism. g also has the property that for all points $p \to Z$, the fiber $g^{-1}(p) = p \underset{Z}{\times} X$ is connected.

Proof: $Z \to \operatorname{Spec} f_* \mathcal{O}_X$ and f is proper. By the finiteness theorem IV.4.1, $f_* \mathcal{O}_X$ is a coherent $\mathcal{O}_Y$-module so h is finite. Applying I.1.21, g is proper, as well as a Stein map.

Since $g_* \mathcal{O}_X = \mathcal{O}_Z$, and g is proper, g must be surjective, so the fiber of g over a point p in Z is nonempty.

Let p be a point of Z, and Z_1 the atom of Z at p (II.6.13). The map $Z_1 \to Z$ is flat (II.6.15) so if we write $X_1 = Z_1 \underset{Z}{\times} X$, then the associated pullback $g_1:Z_1 \to X_1$ is again Stein (II.4.17), and of course still proper. p is a closed point of Z_1 and $g_1^{-1}(p) = g^{-1}(p)$. Thus we may assume in particular that Z is affine and p is a closed point of Z. Let us write $Z = \operatorname{Spec} A$ and let $m \subset A$ be the maximal ideal defining p.

Let F be any coherent sheaf on X. By the holomorphic functions theorem 3.1, there is an isomorphism

$$\varphi_n : (R^n f_*(F))^\wedge_p \to \varprojlim_k H^n(f^{-1}(p), F \otimes_{\mathcal{O}_Z} \mathcal{O}_{Z/m^k}).$$

We apply this in the case $F = \mathcal{O}_X$, $n = 0$.

Then $(R^n f_*(F))^\wedge_p = \Gamma(X, f_* \mathcal{O}_X))^\wedge_p = \Gamma(Z, \mathcal{O}_Z)^\wedge_p = \hat{A}_p$, which is a local ring.

The kth stage of the range of φ_n above is $\Gamma(f^{-1}(p), \mathcal{O}_X \otimes_{\mathcal{O}_Z} \mathcal{O}_{Z/m^k})$. If we write $X_k = f^{-1}(\mathrm{Spec}\, \mathcal{O}_{Z/m^k})$, this k^{th} stage is $\Gamma(X_k, \mathcal{O}_{X_k})$.

Suppose X_1 is not connected. Then we can write $X_1 = U_1 \sqcup V_1$, a disjoint union, and $\Gamma(U_1, \mathcal{O}_{U_1}) = \Gamma(V_1, \mathcal{O}_{V_1})$. X_k has has the same underlying topological space as X_1 so there exists a decomposition $X_k = U_k \cup V_k$, a disjoint union, with $U_1 = U_k \times_{X_k} X_1$ and $V_1 = V_k \times_{X_k} X_1$. Thus the induced decomposition $\Gamma(X_k, \mathcal{O}_{X_k}) = \Gamma(U_k, \mathcal{O}_{U_k}) \times \Gamma(V_k, \mathcal{O}_{V_k})$ is compatible with the map $\Gamma(X_k, \mathcal{O}_{X_k}) \to \Gamma(X_1, \mathcal{O}_{X_1})$. Hence

$$\varprojlim \Gamma(X_k, \mathcal{O}_{X_k}) = \varprojlim \Gamma(U_k, \mathcal{O}_{U_k}) \times \varprojlim \Gamma(V_k, \mathcal{O}_{V_k})$$

(since inverse limits commute).

Neither of $\varprojlim \Gamma(U_k, \mathcal{O}_{U_k})$ and $\varprojlim \Gamma(V_k, \mathcal{O}_{V_k})$ is trivial (each contains at least the limits of the unit sections), so $\varprojlim \Gamma(X_k, \mathcal{O}_{X_k})$ is a product of rings, hence not a local ring. This is a contradiction. ∎

Another result proved in this context in EGA is Zariski's Main

Theorem. As indicated below, the extension of this theorem to the category of algebraic spaces is trivial.

Theorem 4.2: (Zariski's Main Theorem) Let Y be a noetherian locally separated algebraic space and $f:X \to Y$ a quasiprojective map. Then there exist open subspaces X' of X and Y' of Y, with $f(X') \subseteq Y'$, such that

i) $f\big|_{X'} : X' \to Y'$ is finite

and ii) For each point $x \in X$, x is in X' if and only if x is isolated in its fiber $f^{-1}(f(x))$.

Proof: The assertion is local on Y so we can assume that Y is affine. Then X must be a scheme (II.7.6) so this follows from the equivalent assertion for schemes EGA III.4.4.3. ∎

In the rest of this section, we apply these two results in the case of varieties.

Definition 4.3: Let k be an algebraically closed field fixed throughout the discussion. A prevariety X is a reduced algebraic space X given with a map $f:X \to \operatorname{Spec} k$ (the structure map) which is separated and of finite type. X is quasiprojective if f is quasiprojective; affine if f is affine. X is a variety if f is proper and X is also irreducible.

If X is a prevariety, it must be noetherian (II.3.7) so there is an etale covering $\pi:Y \to X$ with Y an affine prevariety, the spectrum of a noetherian ring, and π quasifinite.

Given a point p in a prevariety X, the <u>dimension of p</u> is the dimension of the complete local ring (the atom) of X at p. For any stale covering $\pi: Y \to X$, the point q in Y with $\pi(q) = p$, dim p = dim q and this is the dimension in the usual sense (EGA 0_{IV}.16.1) and is an integer.

The <u>dimension of X</u> is the maximum, $\dim X = \max_{p \in X} \dim p$. If X has dimension 1, X is called a <u>curve</u>; if dimension 2, a <u>surface</u>.

If dim X = n and X is a variety, then the generic point of X is the unique point of dimension n. In this case, given another point $p \in X$ with dim p = m, we say the <u>codimension of p in X</u> is the integer n - m.

In II.6.8, we showed that a prevariety X has a dense open subspace $X_1 \subset X$ which is a scheme. In particular, the generic point x_0 of X must be in this dense subset, so x_0 is schemelike.

<u>Theorem 4.4:</u> Let X be a normal variety, and $x \in X$ a point of codimension one. Then x is schemelike.

Before giving the proof, we need the following lemmas.

<u>Lemma 4.5:</u> Let $f: Y \to X$ be a map of prevarieties and $p \in Y$ be any point with q = f(p). Then dim q $\leq$ dim p.

<u>Proof:</u> Since the notion of dimension is local, we can assume both X and Y are affine prevarieties. This is then EGA 0_{IV}.16.1.5. ∎

<u>Lemma 4.6:</u> Let Y and X be varieties and $f: Y \to X$ be a birational

surjective projective morphism. Then dim X = dim Y. Let $p \in Y$ be any point with $q = f(p)$ its image in X. Then codim $q \geq$ codim p.

Proof: An immediate corollary of Lemma 4.5. ∎

Lemma 4.7: Let $f:Y \to X$ be a finite birational map of varieties. Suppose X is normal. Then f is an isomorphism.

Proof: The assertion is local on X, so we can assume that X is affine. Since f is affine, Y is also affine. In the case of schemes, this is proved in EGA III.4.4.9 as a corollary of the Zariski Main Theorem. ∎

Lemma 4.8: Let X be a projective variety and $A \subset X$ a nonempty subset of points such that all points of A are of codimension one and A is connected in the induced topology. Then either A consists of just one point, or the closure of A is all of X.

Proof: The closure of A, $\bar{A}$, is noetherian and hence has only a finite number of components. If $\bar{A}$ is not all of X, the generic points of these components are the only points of codimension one contained in $\bar{A}$, so A is finite. If A is finite and connected (and X is separated) A must have just one point, or none. ∎

Proof of theorem 4.4: Applying Chow's Lemma IV.3.1, there is a projective surjective birational map $f:Y \to X$ with Y projective over k(so in particular a scheme). A priori, Y may not be reduced. But since $Y \to X$ is birational, and X is reduced, there is a dense open reduced subset of Y. Hence the map $Y_{red} \to Y$ is birational. This map is also clearly projective and surjective so we may as well have at the first considered Y_{red}. Hence we can assume Y is reduced.

Y can also be chosen to be irreducible. For if $Y \to Y_1 \cup Y_2$, with Y_1 and Y_2 closed subsets of Y, then since $Y \to X$ is surjective, there must be a point p in Y_1 (or in Y_2) whose image is the generic point of X. Then $Y_1 \to X$ (or $Y_2 \to X$) is surjective, projective and birational.

Hence Y is a projective variety. By lemma 4.6, dim Y = dim X. Clearly the generic point y_0 of Y is the unique point of Y mapping to the generic point x_0 of X. Now let $x \in X$ be a point of codimension one and $y \in Y$ a point with $f(y) = x$. By lemma 4.6, y must be the generic point of Y or a point of codimension one. But since $x \neq x_0$, $y \neq y_0$, so y is of codimension one. Also, since f is surjective, the closure of $f^{-1}(x)$ (which is contained in f^{-1}(the closure of x)) is a closed subspace of Y but cannot be all of Y. The Connectedness Theorem 4.1 implies that $f^{-1}(x)$ is connected. Hence by lemma 4.8, $f^{-1}(x)$ consists of a single point y. In particular this point y is isolated in its fiber.

We now apply the Zariski Main Theorem. There is an open subspace $X_1 \subset X$ and an open subspace $Y_1 \subset Y$ such that if $f(Y_1) \subset X_1$, the map $f_1 : Y_1 \to X_1$ is finite and Y_1 contains all points isolated in their fibers under f. Our map f is a birational map of irreducible varieties, so f_1 is birational. Y is normal so Y_1 is normal. Applying lemma 4.7, f_1 must be an isomorphism.

If $x \in X$ is of codimension one, then as argued above, x must be in X_1. Y_1 is a scheme, since it is an open subspace of a projective variety, so $Y_1 = X_1$ is an open neighborhood of x which is a scheme. ∎

<u>Theorem 4.9:</u> Let X be a curve over an algebraically closed field k. Then X is a scheme.

<u>Proof:</u> By Chow's Lemma, there is a prevariety Y and map $\pi : Y \to X$ which is projective, surjective and birational. As in the previous theorem 4.4, we can assume Y is also reduced irreducible and of dimension one. Also, π is clearly quasifinite. Since π is quasifinite and proper, it must be finite. (This statement is local on X, and if X is affine, by II.6.15, Y must be a scheme, and we can apply the corresponding proposition EGA III.4.4.2 for schemes.

Let $U \to X$ be the subset of X for which $\pi^{-1}(U) \to U$ is an isomorphism. Let $p \in \pi^{-1}(U)$ and consider the restrictions $Y - y \to X - \pi(y)$. This is again finite. But since the scheme Y is a curve, any open subspace $Y - y$ of Y must be affine. (This is a well-known fact. One can prove it by, for example, applying the Nakai criterion to show that $Y - y$ is quasiprojective and then noting that it is obviously true for projective varieties.) Applying the Chevalley Criterion (III.4.1) $X - \pi(y)$ is affine, hence a scheme. Since $\pi(y) \in X$ is a schemelike point, we are done (II.6.6).

The reader will note this theorem is easily extendible to the case where X is an arbitrary prevariety of dimension one.

4.10: Without making specific definitions, we note that the thoerem above can be interpreted as saying that Weil divisors on a normal variety are all defined by schemelike points.

Also, the usual proofs [XVa] showing in the category of schemes that nonsingular surfaces are always projective carries over to the case of algebraic spaces. One should not be too optimistic however. Recall we gave examples in the Introduction of a singular surface and a nonsingular three-fold neither of which were schemes.

5. Completions of Modules of Homomorphisms

Definition 5.1: Let X be a noetherian separated algebraic space and F and G quasicoherent $\mathcal{O}_X$-modules. The hom-sheaf of F and G on X, $\underset{\sim}{\mathrm{Hom}}_{\mathcal{O}_X}(F,G)$ is defined by, for every $Y \to X$ etale, $\Gamma(Y, \underset{\sim}{\mathrm{Hom}}_{\mathcal{O}_X}(F,G)) = \mathrm{Hom}_{\mathcal{O}_Y}(F|_Y, G|_Y)$. This is a sheaf and if F and G are coherent, so is $\underset{\sim}{\mathrm{Hom}}_{\mathcal{O}_X}(F,G)$. The category of quasicoherent sheaves on X has enough injectives so we can define derived functors $\underset{\sim}{\mathrm{Ext}}^q_{\mathcal{O}_X}(F,-)$, $q \geq 0$, of the functor $\underset{\sim}{\mathrm{Hom}}_{\mathcal{O}_X}(F,-)$.

Of course, we can also define the derived functors $\mathrm{Ext}^n_{\mathcal{O}_X}(F,G)$ of the functor $\mathrm{Hom}_{\mathcal{O}_X}(F,G)$ of the functor $\mathrm{Hom}_X(F,G)$. These are two compared in the following proposition.

Proposition 5.2: With the notation above, there is a spectral sequence E(F,G)

$$E_2^{p,q} = H^p(X, \underset{\sim}{\mathrm{Ext}}^q(F,G)) \Rightarrow \mathrm{Ext}^{p+q}(F,G)$$

Proof: We need to show that for all quasicoherent sheaves G on X, there is an injective quasicoherent $\mathcal{O}_X$-module I and an injection $G \to I$ such that $H^p(X, \underset{\sim}{\mathrm{Hom}}(F,I)) = 0$, $p > 0$.

Let $\pi : Y \to X$ be an etale covering of X. Since X is noetherian and separated, we can assume that Y is affine and π is affine. Then as in I.4.9, we can find an injective quasicoherent $\mathcal{O}_Y$-module J with $\pi_* J$ injective, and an immersion $G \to \pi_* J$. Hence we are reduced to showing that for all quasicoherent modules H on Y,

$H^p(X, \underset{\sim}{Hom}_{\mathcal{O}_X}(F,\pi_*H)) = 0,\ p > 0.$

But $\pi^* \underset{\sim}{Hom}_{\mathcal{O}_Y}(\pi^*F,H) = \underset{\sim}{Hom}_{\mathcal{O}_X}(F,\pi_*H)$. (This assertion is local on X and for X affine is the usual adjointness relation.)

Hence $H^p(X, \underset{\sim}{Hom}_{\mathcal{O}_X}(F,\pi_*H)) = H^p(X,\pi_*\underset{\sim}{Hom}_{\mathcal{O}_Y}(\pi^*F,H))$. Since π is affine, this is zero (I.4.16). ∎

If $\mathfrak{X}$ is a formal algebraic space, we can write the same definition and the proposition 5.2 carries over easily to this case.

5.3: We fix the following notation for the rest of this section. A is a noetherian ring, I an ideal of A, X an algebraic space, separated and finite type over Spec A, F and G are two coherent $\mathcal{O}_X$-modules the intersection of whose supports is proper over Spec A. $X_0 = X \times_{\text{Spec } A} \text{Spec } (A/I)$ and $\hat{X}$ is the completion of X along X_0. $i:\hat{X} \to X$ is the canonical map. Recall $\hat{F} = i^*F$, $\hat{G} = i^*G$. (Prop. 2.3)

Lemma 5.4:

i) There exists a canonical isomorphism $i^*(\underset{\sim}{Ext}^q_{\mathcal{O}_X}(F,G)) \xrightarrow{\sim} \underset{\sim}{Ext}^q_{\mathcal{O}_{\hat{X}}}(\hat{F},\hat{G}),\ q \geq 0.$

ii) There exists a canonical map of spectral sequences $E(F,G) \to E(\hat{F},\hat{G})$ which in the $E_2^{p,q}$ terms is the map $H^p(X, \underset{\sim}{Ext}^q_{\mathcal{O}_X}(F,G)) \to H^p(\hat{X}, \underset{\sim}{Ext}^q_{\mathcal{O}_{\hat{X}}}(\hat{F},\hat{G}))$ compatible with the above isomorphism.

Proof: This is a homological algebra calculation identical to that of EGA 0_{III}.12.3.4, 5 for flat maps of ringed spaces. ∎

<u>Proposition 5.5:</u> For all $n \geq 0$, $\underset{\mathcal{O}_X}{\mathrm{Ext}^n}(F,G)$ is an A-module of finite type and its I-adic completion is canonically isomorphic to $\mathrm{Ext}^n_{\mathcal{O}_{\hat{X}}}(\hat{F},\hat{G})$.

<u>Proof:</u> Consider the spectral sequence $E(F,G)$ of Prop. 5.2. As mentioned above, $\underset{\sim}{\mathrm{Ext}}^q_{\mathcal{O}_X}(F,G)$ is a coherent module. Its support is contained in the intersection of the supports of F and G and hence is proper over Spec A. Hence by the Finiteness Theorem IV.4.1, the $E_2^{p,q}$ terms of $E(F,G)$ are A-modules of finite type. This implies that all of the $E_r^{p,q}$ terms are of finite type so the abutment $\mathrm{Ext}^n_{\mathcal{O}_X}(F,G)$ is of finite type.

Since $\underset{\sim}{\mathrm{Ext}}^q_{\mathcal{O}_X}(F,G)$ is of finite type, there is an isomorphism $\underset{\sim}{\mathrm{Ext}}^q_{\mathcal{O}_X}(F,G)^\wedge \cong i^*(\underset{\sim}{\mathrm{Ext}}^q_{\mathcal{O}_X}(F,G))$ by Prop. 2.3. Combining this with the isomorphism of Lemma 5.4(i) we have an isomorphism $\underset{\sim}{\mathrm{Ext}}^q_{\mathcal{O}_X}(F,G)^\wedge \cong \underset{\sim}{\mathrm{Ext}}^q_{\mathcal{O}_{\hat{X}}}(\hat{F},\hat{G})$. Applying the Holomorphic Functions theorem 3.1, there is, for each $p \geq 0$, a canonical isomorphism $H^p(\hat{X}, \underset{\sim}{\mathrm{Ext}}^q_{\mathcal{O}_{\hat{X}}}(\hat{F},\hat{G})) \cong H^p(X, \underset{\sim}{\mathrm{Ext}}^q_{\mathcal{O}_X}(F,G))^\wedge$. Thus in the spectral sequence $E(\hat{F},\hat{G})$, we have $E_2^{p,q}(\hat{F},\hat{G}) = E_2^{p,q}(F,G)^\wedge = E_2^{p,q}(F,G) \underset{A}{\otimes} \hat{A}$ where $\hat{A}$ is the I-adic completion of A.

Now consider the map $E(F,G) \to E(\hat{F},\hat{G})$ of Lemma 5.4(ii). $\hat{A}$ is a flat A-module so we can apply the functor $\underset{A}{\otimes} \hat{A}$ to all the terms of $E(F,G)$ to get a new spectral sequence. Since all the terms of $E(F,G)$ are A-modules of finite type, the functor $\underset{A}{\otimes} \hat{A}$ is identical here to the completion functor and this new spectral sequence can be denoted $E(F,G)^{\wedge}$.

All the terms of $E(\hat{F},\hat{G})$ are A-modules of finite type so are I-adically complete. Hence the universal mapping properties of the completion functor gives a map $E(F,G)^{\wedge} \to E(\hat{F},\hat{G})$. In the $E_2^{p,q}$ terms this map is

$$H^p(X, \underset{\sim}{\mathrm{Ext}}^q_{\mathcal{O}_X}(F,G))^{\wedge} \to H^p(X, \underset{\sim}{\mathrm{Ext}}^q_{\mathcal{O}_{\hat{X}}}(\hat{F},\hat{G}))$$

which was shown above to be an isomorphism. Hence the map on the abutments $\mathrm{Ext}^n_{\mathcal{O}_X}(F,G)^{\wedge} \to \mathrm{Ext}^n_{\mathcal{O}^{\wedge}_X}(\hat{F},\hat{G})$ is an isomorphism. ∎

6. The Grothendieck Existence Theorem

Let A be a noetherian adic ring with ideal of definition I and let Y = Spec A, Y' = Spec A/I and $\hat{Y}$ = Spf A, (the completion of Y along Y'). Let $f:X \to Y$ be a morphism of finite type with X an algebraic space, and $\hat{X}$ be the completion of X along $X' = Y' \times_Y X$. Let $\hat{f}:\hat{X} \to \hat{Y}$ be the extension of f. Finally let F be a coherent sheaf on X and $\hat{F}$ its completion on $\hat{X}$.

Proposition 6.1: In this situation, suppose the support of F is proper over Y. Then the canonical morphisms

$$P_i : H^i(X,F) \to H^i(\hat{X},\hat{F})$$

are isomorphisms.

Proof: The A-modules $H^i(X,F)$ are modules of finite type over A by the finiteness theorem (IV.4.1) so are equal to their separable completions in the I-adic topology. The proposition is then a special case of the Holomorphic Functions Theorem (3.1). ∎

Proposition 6.2: Let F and G be two coherent $\mathcal{O}_X$-modules such that the intersection of their supports is proper over Y. Then the canonical map

$$\mathrm{Hom}_{\mathcal{O}_X}(F,G) \to \mathrm{Hom}_{\hat{X}}(\hat{F},\hat{G})$$

$$u \leadsto \hat{u}$$

is an isomorphism. Further, if f is closed, then $\hat{u}$ is injective (respectively surjective) iff u is injective (surjective).

Proof: The first assertion is a particular case of 5.5 since $\mathrm{Hom}_{\mathcal{O}_X}(F,G)$ is an A-module of finite type, so equal to its separable completion. To see the second, we first claim that $\hat{u}$ is injective (surjective) iff there is an open set $V \to X$ such that $X' \subset V$ and u is injective (surjective) on V. This is true since the kernel K and cokernel Q are coherent. If $\hat{K}$, say, vanishes, then K = IK so the support of K, a closed set, is disjoint from V(I) = X'. Then X - V(I) is an open neighborhood of X on which u is injective.

However, if f is closed any such V must be identical to X. To prove this, let C be the complement of V in X. Then $C \cap X' = \emptyset$ and since $X' = f^{-1}(Y')$, $f(C) \cap Y' = \emptyset$. But f(C) is closed and every nonempty closed set of Y meets Y' since Y = Spec A is separable in its I-adic topology so I is contained in the radical of A and Y' = Spec A/I. Hence f(C), so C, is empty.

Thus $\hat{u}$ is injective (surjective) iff u is. ∎

Thus we have a functor $F \rightsquigarrow \hat{F}$ mapping the category of coherent $\mathcal{O}_X$-modules whose support is proper over Y to coherent $\mathcal{O}_{\hat{X}}$-modules with support proper over $\hat{Y}$. By the above lemma, this functor establishes an equivalence of the first category with a full subcategory of the second.

<u>Theorem 6.3:</u> (The Grothendieck Existence Theorem, Cf. EGA III.5.1.4). Let A be an adic noetherian ring, Y = Spec A, I an ideal of A, Y' = V(I), $f:X \to Y$ a separated finite type morphism and X any algebraic space, $X' = f^{-1}(Y') = Y' \times_Y X$, $\hat{Y} = \text{Spf } A = \hat{Y}/_{Y'}$, $\hat{X} = \hat{Y} \times_Y X = \hat{X}/_{X'}$, $\hat{f}:\hat{X} \to \hat{Y}$ the completion of f. Then the functor $F \rightsquigarrow \hat{F}$ is an equivalence of the category of coherent $\mathcal{O}_X$-modules with support proper over Spec A, with the category of coherent $\mathcal{O}_{\hat{X}}$-modules with support proper over Spf A.

The most important case is the

<u>Corollary 6.4:</u> Suppose X is proper over Y. Then the functor $F \rightsquigarrow \hat{F}$ is an equivalence of the categories of coherent $\mathcal{O}_X$-modules and coherent $\mathcal{O}_{\hat{X}}$-modules. ∎

<u>Proof of 6.3:</u> In this proof we will say a coherent $\mathcal{O}_{\hat{X}}$-module is <u>algebrizable</u> if it is isomorphic to a completion $\hat{F}$ of a coherent $\mathcal{O}_X$-module F with support proper over Y. We first prove Lemmas 6.5-6.9.

<u>Lemma 6.5:</u> Let F' and G' be two algebrizable $\mathcal{O}_{\hat{X}}$-modules. Then for any homomorphism $u:F' \to G'$, Ker (u), Im(u), and Coker(u) are algebrizable.

<u>Proof:</u> Let $F' = \hat{F}$, $G' = \hat{G}$ where F and G are coherent $\mathcal{O}_X$-modules with proper support. Then $u:\hat{F} \to \hat{G}$ is of the form $\hat{v}:\hat{F} \to \hat{G}$ where

$v:F \to G$ is some map by 6.2. $\mathrm{Ker}(\hat{v})$ is isomorphic to $(\mathrm{Ker}(v))^{\wedge}$ since $F \rightsquigarrow \hat{F}$ is exact and the support of $\mathrm{Ker}(v)$ is closed subspace of $\mathrm{Supp}(F)$ and so is proper over Y. Hence $\mathrm{Ker}(u)$ is algebrizable. Similarly for $\mathrm{Im}(u)$ and $\mathrm{Coker}(u)$. ∎

<u>Lemma 6.6:</u> Let $0 \to H \to F \to G \to 0$ be an exact sequence of coherent $\mathcal{O}_{\hat{X}}$-modules such that G and H are algebrizable. Then F is algebrizable.

<u>Proof:</u> Suppose $H = \hat{B}$ and $G = \hat{C}$ for coherent $\mathcal{O}_X$-modules B and C with proper support. Then F defines an element of $\mathrm{Ext}^1_{\mathcal{O}_{\hat{X}}}(\hat{B},\hat{C})$ which by 5.5 is identical to $\mathrm{Ext}^1_{\mathcal{O}_X}(B,C)$. Let A be the coherent $\mathcal{O}_X$-module (determined up to isomorphism) representing this element of $\mathrm{Ext}^1_{\mathcal{O}_X}(B,C)$. Then $\mathrm{Supp}\ A \subset (\mathrm{Supp}\ B \cap \mathrm{Supp}\ C)$ so is proper, and $\hat{A}$ is isomorphic to F. ∎

<u>Corollary 6.7:</u> Let $u:F \to G$ be a homomorphism of coherent $\mathcal{O}_{\hat{X}}$-modules. Then if G, $\mathrm{Ker}(u)$ and $\mathrm{Coker}(u)$ are algebrizable, so is F.

<u>Proof:</u> Immediate from 6.5 and 6.6. ∎

<u>Lemma 6.8:</u> Let $h:Z \to Y$ be a morphism of finite type and $\hat{Z}$ the completion of Z along $Z' = Y' \times_Y Z$. Let $g:Z \to X$ be a proper Y-morphism and $\hat{g}:\hat{Z} \to \hat{X}$ its completion. Then for every algebrizable $\mathcal{O}_{\hat{Z}}$-module F', $\hat{g}_*(F')$ is an algebrizable $\mathcal{O}_{\hat{X}}$-module.

<u>Proof:</u> Let F be a coherent $\mathcal{O}_Z$-module such that $F' = \hat{F}$. Then by 3.1, $\hat{g}_*(\hat{F}) \cong g_*(F)^{\wedge}$. ∎

<u>Lemma 6.9:</u> Let X be a separated noetherian algebraic space, X' a closed subset of X, $f:Z \to X$ a proper morphism, $Z' = f^{-1}(X')$, $\hat{X} = X/_{X'}$, $\hat{Z} = Z \times_X \hat{X} = Z/_{Z'}$, and $\hat{f}:\hat{Z} \to \hat{X}$ the completion of f. Let M be a coherent ideal of $\mathcal{O}_X$ such that if U is the open complement of Supp $\mathcal{O}_{X/M}$ in X, the restriction $f^{-1}(U) \to U$ of f is an isomorphism. Then for every coherent $\mathcal{O}_{\hat{X}}$-module F, there is an integer $n > 0$ such that the kernel and cokernel of the canonical map $\rho_F:F \to \hat{f}_*\hat{f}^*F$ is annihilated by $\hat{M}^n$.

<u>Proof:</u> The statement is local on X so we can assume X is affine. Say X = Spec B, and X' = V(I) for some ideal I of B.

Also we can assume B is adique and I is an ideal of definition of B. To see this, let B_1 be the separable completion of B in the I-adic topology and $I_1 = IB_1$. Then B_1 is an adique noetherian ring with ideal of definition I_1. Put $X_1 = \operatorname{Spec} B_1$ and $h:X_1 \to X$ the canonical map. Then $X_1' = h^{-1}(X')$ is identical to $V(I_1)$. Put finally $Z_1 = Z \times_X X_1$, $f_1:Z_1 \to X_1$ the induced map, which is proper, and let $\hat{X}_1$ be the completion of X_1 along X_1', $\hat{Z}_1 = Z_1 \times_{X_1} \hat{X}_1$ the completion of Z_1 along $Z_1' = f_1^{-1}(X_1')$ and $\hat{f}_1$ the completion of f. Then $\hat{h}:\hat{X}_1 \to \hat{X}$ is an isomorphism so $\hat{Z}_1 \to \hat{Z}$ is an isomorphism. Hence $\hat{f}_1 = \hat{f}$.

Finally, $M_1 = h^*(M)$ is a coherent ideal of $\mathcal{O}_{X_1}$ and

$\mathrm{Supp}(\mathcal{O}_{X_1/M_1}) = h^{-1}(\mathrm{supp}(\mathcal{O}_X(M))$ so, letting U_1 be the complement of $\mathrm{Supp}(\mathcal{O}_{X_1/M_1})$ in X_1, $U_1 = h^{-1}(U)$ whence the restriction $f_1^{-1}(U_1) \to U_1$ of f_1 is an isomorphism. Also $\hat{M}$ and $\hat{M}_1$ are identified by $\hat{h}$. All the hypotheses of the lemma are then satisfied by X_1, $\hat{X}_1$, f_1 and M_1 and one can then assume B is an adique noetherian ring and I an ideal of definition of B.

We have $\hat{X} = \mathrm{Spf}\, B$ and a given sheaf F on $\hat{X}$ is hence of the form $\hat{G}$ where G is a coherent 0_X-module. By 2.3, $\hat{f}^*(F) = (f^*(G))^{\wedge}$ and by 3.1, $\hat{f}_*((f^*(G))^{\wedge})$ is canonically equal to $(f_*(f^*(G)))^{\wedge}$ so ρ_F is the completion $\hat{\rho}_g$ where $\rho_g : G \to f_* f^*(G)$. But the kernel P and cokernel Q of ρ_g are coherent and their restrictions to U are zero. Hence (using II.5.8) there is an integer $n > 0$ such that $M^n P = M^n R = 0$ so $\hat{M}^n \hat{P} = \hat{M}^n \hat{R} = 0$. ∎

<u>Final Proof of 6.3:</u> For the case $f:X \to Y$ <u>quasi-projective</u>, we canmerely quote [EGA III.5.2] since Y is affine so X is a scheme and this is the case treated there.

For the general case we use noetherian induction and assume the theorem is true for every proper closed subspace T of X. (We take the completion $\hat{T}$ to be the completion of T along $T' = T \cap X'$). We can assume $X \neq \emptyset$.

Since f is separated and of finite type, Chow's Lemma (IV.3.1) applies and there is a Y-space Z and a Y-morphism $g:Z \to X$ such that the composition $Z \to Y$ is quasiprojective and g is projective, surjective, and there is an open subset U of X such that $g^{-1}(U) \to U$ is an isomorphism.

Let M be a coherent ideal of $\mathcal{O}_X$ defining the closed subspace X - U and F a coherent $\mathcal{O}_{\hat{X}}$-module with support E proper over Y. Let $\hat{Z}$ be the completion of Z along $h^{-1}(Y')$ and $\hat{g}:\hat{Z} \to \hat{X}$ the completion of g. Then $\hat{g}^*(F)$ is a coherent $\mathcal{O}_{\hat{Z}}$-module whose support is contained in $g^{-1}(E)$. Thus the support of $\hat{g}^*(F)$ is proper over Y. As h is quasiprojective $\hat{g}^*(F)$ is algebrizable. Hence $\hat{g}_*\hat{g}^*(F)$ is an algebrizable $\mathcal{O}_{\hat{X}}$-module since g is proper, by 6.8. Applying 6.9, the kernel P and cokernel R of the canonical map $\rho_F:F \to \hat{g}_*(\hat{g}^*(F))$ are annihilated by some power $\hat{M}^n$. Let T be the closed subspace of X defined by M^n, $T = \text{Spec}\, \mathcal{O}_X/M^n$, and $j:T \to X$ the canonical injection. One can write $P = \hat{j}_*(\hat{j}^*(P))$ and $R \simeq \hat{j}_*(\hat{j}^*(R))$. Since U is nonempty the induction hypothesis implies that $\hat{j}^*(P)$ and $\hat{j}^*(R)$ are algebrizable so by 6.8, P and R are algebrizable. Then by 6.8, F is algebrizable. ∎

INDEX

affine etale covering 103
affine formal scheme 211
affine image of a map 124
algebraic space 92
- -, formal 215
- -, integral 144
- -, irreducible 127
- -, affine 104
- - is a scheme 104
- -, locally noetherian 105
- -, locally separated 97
- -, n-dimensional 106
- -, noetherian 105
- -, nonsingular 106
- -, normal 106
- -, quasicompact 105
- -, reduced 106
- -, separated 97
algebrizable $_X$-module 247
ample invertible sheaf 142
atom 133
Axiom A_0 30
Axioms $S_1, S_2, S_3(a), S_3(b)$ 36

blowing up 19

Chevalley's Theorem 169
Chow's Lemma 190
closed subcategory 32
closure, geometric 124
closure, scheme-theoretic 48
coherent sheaf 41
cohomology of sheaves (quasicoherent) 117
Completeness Lemma 155
completion of a ring along an ideal 210
completion of an algebraic space along a closed subspace 216
components of a space 145
Connectedness Theorem 233
covering, affine etale 103
covering map 30
covering, representable etale 92
curve 236

decomposition into components 145
dense, geometrically 124
- -, topologically 125
descent, effective 32
descent theory for modules 54
- -over adic rings 207
Devissage 173
diagonal component 6, 80
dimension, codimension of a point 236
divisor, Cartier 149
- -, Weil 240

effective descent 32, 34
equivalence relation, categorical 72
- -, effective 72
- -, "finitely presented" 83
- -, induced 72
- -, quotient of 72
- -, on algebraic spaces 113
etale covering, representable 92

etale topology, affine formal schemes 211
- -, algebraic spaces 102
- -, schemes 59
Extension Lemma 157

fiber of a quasicoherent sheaf at a point 148
filtration, I-good 220
Finiteness Theorem 202
flat topology, affine schemes 58
- -, schemes 58
formal scheme, affine 211
- -, as formal algebraic space 216
formal algebraic space 215
general definitions 215
ideal of definition 217
function field 148
function field sheaf 148

generic point 144
generic rank of a coherent sheaf 150
geometric closure 124
geometrically dense 124
gluing data 5
graded ring 218
graded R-module 219
Grothendieck Existence Theorem 247
Grothendieck topology 29
group actions on an algebraic space 177
- -, free action 178
geometric quotient of - 178
fixed point locus 186

Hilbert Basis Theorem 108
Hilbert Scheme 2
Holomorphic Functions Theorem 227

immersion 42
- -, closed 42
- -, open 42
inductive property of algebraic spaces 128
invertible sheaf 42

Jacobian condition 59

local construction 33
- -, effective 33
local on the domain 32,34
local ringed space 39
locally separated 97

map = morphism
map of adic rings 205
- -, adic 205
- -, faithfully flat 207
- -, flat 207
- -, formally etale 208
- -, formal etale covering 208
- -, of finite type 207
map of algebraic spaces 92
- -, affine 108
- -, birational 144
- -, closed 133
- -, closed immersion 108
- -, covering map 101
- -, etale 101
- -, etale surjective 101
- -, finite 139
- -, faithfully flat 107
- -, flat 107
- -, immersion 108
- -, locally of finite type 107
- -, locally of finite presentation 107
- -, locally quasifinite 107

map of algebraic spaces, cont.
- -, locally separates 109
- -, of finite presentation 107
- -, of finite type 107
- -, open immersion 108
- -, projective 141
- -, quasiaffine 108
- -, quasicompact 105
- -, quasifinite 107
- -, quasiprojective 141
- -, quasiseparated 109
- -, reduced closed immersion 109
- -, section of 110
- -, separated 109
- -, Stein 123
- -, surjective 107
- -, universally closed 133
- -, universally open 107
map of affine formal schemes 211
- -, imbedding 211
- -, closed imbedding 211
- -, open imbedding 211
map of formal algebraic spaces, general definitions 215, 216
- -, proper 218
map of local ringed spaces 39
map of rings, flat 53
- -, faithfully flat 53
map of schemes 40
- -, affine 47
- -, bijective 43
- -, closed 43
- -, etale 59
- -, faithfully flat (fflat) 56
- -, finite 50
- -, flat 56
- -, formally etale 61
- -, injective 43
- -, locally of finite presentation 46
map of schemes, cont.
- -, locally of finite type 46
- -, of finite presentation 50
- -, of finite type 50
- -, open 43
- -, quasiaffine 47
- -, quasicompact 45
- -, quasifinite 47, 50
- -, quasiseparated 46
- -, radiciel 43
- -, separated 44
- -, surjective 43
- -, union of Zariski open sets 44
- -, universally closed 43
- -, universally bijective 43
- -, universally injective 43
- -, universally open 43
- -, unramified 61
- -, Z-open 44
Mittag-Leffler condition 221
modification 19
module over radic ring, continuous 205
- -, complete tensor product 206
- -, complete Hom set 206
Moisezon Space 23
morphism = map

Noetherian Induction 128

"open subspace where X is a scheme" 131

Picard Group 150
points 129
- -, equivalence of 129
- -, generic 144
- -, geometric 129
- -, residue field at 129

points, schemelike 131
presheaf 30
prime spectrum 38
projective n-space over an
algebraic space 141

quasicompact (scheme) 41
quasicoherent sheaf 42
- -, in the etale topology 67

rank, generic, of a coherent
sheaf 150
rank of a locally free sheaf 42
representable etale covering 92
ring, adic 204
- -, discrete adic 205
- -, associated graded 219
- -, graded 218

scheme 41
- -, affine 40
- -, locally noetherian 44
- -, noetherian 49
- -, nonsingular 45
- -, of dimension n over a
ground field 45
- -, reduced 45
- -, separated 44
section of a map 110
separated algebraic space 97
Serre Criterion 163
- -, Weak 161
Serre Finiteness Theorem 142
sheaf 30
- -, abelian 30
- -, coherent 41
- -, Hom and Ext 241
- -, invertible 42
- -, quasicoherent 42
- -, structure 104
- -, torsion 151
- -, torsion-free 151
sheaf on an algebraic space,
coherent 113
- -, locally free of rank r 113
- -, map of 113
- -, quasicoherent 113
sheaf, global, on the category
of affine formal schemes 212
- -, continuity 212
- -, structure 213
Sheaf Criterion for
Isomorphism 121
space, local ringed 39
Spec R 38
spectrum 38
stable (class of maps) 32
- -, (class of objects) 31
- -, (property of maps) 34
- -, (property of objects) 34
stalk 148
stalk, geometric 147
Stein factorization of a map 124
(Also see p. 233)
strict initial object 35
strict projective system 221
subcategory, closed 32
subscheme 41
- -, closed 40
- -, open 40
subspace of an algebraic
space 109
subspace, closed 109
- -, open 109
support of a sheaf on a
scheme 52
- - on an algebraic space 127
surface 236
symmetric powers of
projective spaces 188
- equivalence relation 73
- -, effective 73
- -, quotient 73
topologically dense 125

topology 29
topology associated to a closed subcategory 35
topology, etale, local vs. global 102
- -, affine formal schemes 211
- -, algebraic spaces 102
- -, schemes 63
topology, fppf (schemes) 59
topology, flat (schemes) 55
topology, Zariski, on an algebraic space 132
- -, on schemes 44

truncation, of adic ring 205
- -, of a formal algebraic space 217

universal effectively epimorphic family (UEEF) 34

variety 235
prevariety 235
dimension of 236

Zariski's Main Theorem 235
Zariski topology on schemes 44
on an algebraic space 132

INDEX OF NOTATION

A.m.n. (A,m,n numerals) 3
$X^{\cdot}$ (X a space) 3,91
Spec 47,120
X^h 88
$\mathcal{O}$ 104
$QCS_X, CS_X, MS_X, AS_X, AP_X$ 113
f^*, f^{ab}, f^m 114
$\mathbb{A}^n_X$ 121
X_{red} 126
$|X|$ 132
$\mathbb{P}^n_X$ 141
Spf 216

GENERAL ASSUMPTIONS

All the schemes (and algebraic spaces) considered are always assumed to be quasiseparated. (See I.2.25 and II.1.9) In Chapters IV and V, all algebraic spaces are assumed to be Noetherian and separated.

BIBLIOGRAPHY

EGA J. Dieudonné and A Grothendieck, *Éléments de géométrie algébrique*, Pub. Math. Inst. Hautes Etudes Sci., 1960-

SGA A. Grothendieck, *Séminaire de géométrie algébrique*, 1960-60, Inst. Hautes Etudes Sci. (mimeographed notes).

SGAA M. Artin, A. Grothendieck and J.L. Verdier, *Séminaire de géométrie algébrique*, 1963-64, IHES (mimeographed notes).

SGAD M. Demazure and A. Grothendieck, *Séminaire de géometrie algébrique*, *Schemas en groups*, IHES, (mimeographed notes). (or - Lecture Notes in Mathematics, No. 151-53, Springer-Verlag, 1970)

GT M. Artin, *Grothendieck Topologies*, 1962, Harvard University (mimeographed notes)

I M. Artin, *On the solutions of analytic equations*, Invent.Math. 5 (1968) 277-291

II M. Artin, *Algebraization of formal moduli: I*, in *Global Analysis, Papers in Honor of K. Kodaira*, (D.C.Spencer, S. Iyanaga, Ed.) University of Tokyo Press and Princeton University Press 1970

III M. Artin, *Algebraization of formal moduli: II - Existence of modifications*, Ann.Math., 91, Jan. 1970, 88-135

IV M. Artin, *The implicit function theorem in Algebraic Geometry*, Proc. Bombay Colloquium on Algebraic Geometry, Tata Institute, 1969

V M. Artin, Algebraic Spaces, The Whittemore Lectures, Yale University, 1969, (mimeographed notes)

VI M. Artin, Commutative rings, 1966, Massachusetts Institute of Technology, (mimeographed notes)

VII H. Bass, Algebraic K-Theory. W.A.Benjamin, Inc., New York,1968

VIII N. Bourbaki, Elements de Mathematique, Algebre Commutative, Chapter III, Hermann, Paris, 1961

IX J. Dieudonne, Algebraic Geometry, 1962, Harvard University (mimeographed notes)

X J. Dieudonne, Fondements de la geometrie algebrique modern, 1961, University of Montreal (mimeographed notes)

XI J. Dieudonne, (reprints of the above two),Advances in Mathematics, 3, No. 3, July 1969, 233-414

XII A. Douady, Le probleme des modules pour les sous-espaces analytiques compacts d'un espace analytique donne, Ann. de l'Inst. Fourier, Grenoble, 16, (1966), 1-98

XIII J. Tate, Rational Points on Elliptic Curves, Philips Lectures, Haverford College, April, May 1961

XIV P.Gabriel, Des categories abeliennes, Bull.Soc.Math. France, 90, 1962, 323-448

XV R.Godement, Topologie algebrique et theorie des faisceaux, Hermann, Paris, 1958

XVa J.E.Goodman, Affine open subsets of algebraic varieties and ample divisors, Thesis, Columbia University, 1967

XVI H. Grauert, Uber Modifikationen und exzeptionelle analytische Mengen, Math. Ann. 146 (1962), 331-368

XVII A. Grothendieck, Fondements de la geometrie algebrique, Extraits du Seminaire Bourbaki, 1957-1962, (mimeographed notes)

XVIII H. Hironaka, An example of a non-Kahlerian deformation, Ann. of Math., 75, (1962), 190-

XIX H. Hironaka, Resolution of singularities of an algebraic vareity over a field of characteristic zero, Ann. of Math., 79 (1964) 109-326

XX S. Kleiman, Toward a numerical theory of ampleness, Ann. of Math., 84 (1966), 293-344

XXI T. Matsusaka, Theory of Q-varieties, Pub. Math. Soc. Japan, No. 8, Tokyo, 1965

XXII B.G.Moisezon, On n-dimensional compact varieties with n algebraically independent meromorphic functions, I,II,III Izv.Akad.Nauk SSSR Ser.Mat. 30(1966), 133-174, 345-386, 621-656. Also, Amer. Math. Soc. Translations, ser.2, Vol. 63, 1967

XXIII B.G.Moisezon, Resolution theorems for compact complex spaces with a sufficiently large field of meromorphic functions, Izv.Akad.Nauk SSSR Ser. Mat. 31(1967), 1385. Also Mathematics of the USSR-Izvestis 1 (1967), 1331-1356

XXIV B.G.Moisezon, Algebraic analogue of compact complex spaces with a sufficiently large field of meromorphic functions, I,II,III, Izv. Akad. Nauk SSSR Ser. Mat. 33 (1969) 174-238,323-367

XXV D. Mumford, Introduction to algebraic geometry , preliminary version, 1966, Harvard University, (mimeographed notes)

XXVI D. Mumford, Geometric invariant theory, Ergebnisse der Math., Bd. 34, Springer, Berlin, 1965

XXVII D. Mumford, Picard groups of moduli problems, pp. 33-81 in Arithmetical algebraic geometry (O.F.G.Schilling, Ed.), Harper and Row, New York, 1965

XXVIII J. Nash, Real algebraic manifolds, Ann. of Math., 56 (1952) 405-421

XXIX J.-P. Serre, Faisceaux Algebrique Coherent, Ann. of Math., 61, 1955, 197-278

XXX J.-P. Serre, Geometrie algebrique et geometrie analytique, Ann. Inst. Fourier Grenoble 6 (1956), 1-42

XXXI C.S. Seshadri, Some results on the quotient space by an algebraic group of automorphisms, Math. Ann., 149 (1963) 286-301

XXXII C.L. Siegel, Meromorphe Funktionen auf kompakten analytischen Manningfaltigkeiten, Nachr. Akad. Wiss. Gottingen, Math.-Phys. Kl. IIa, (1955), 71-77

XXXIII M. Raynaud, Passage au Quotient par une Relation d'Equivalence Plate, Proceedings of a Conference on Local Fields, T.A. Springer, Editor, Springer-Verlag, 1967.

XXXIV M. Raynaud, Anneaux Locaux Henseliens, Lecture Notes in Mathematics No. 169, Springer-Verlag, 1970

XXXV P. Deligne and D. Mumford, The Irreducibility of the Space of Curves of Given Genus, Publicationes I.H.E.S., No. 36, 1969

Lecture Notes in Mathematics

Vol. 85: P. Cartier et D. Foata, Problèmes combinatoires de commutation et réarrangements. IV, 88 pages. 1969. DM 8,– / $ 2.20

Vol. 86: Category Theory, Homology Theory and their Applications I. Edited by P. Hilton. VI, 216 pages. 1969. DM 16,– / $ 4.40

Vol. 87: M. Tierney, Categorical Constructions in Stable Homotopy Theory. IV, 65 pages. 1969. DM 6,– / $ 1.70

Vol. 88: Séminaire de Probabilités III. IV, 229 pages. 1969. DM 18,– / $ 5.00

Vol. 89: Probability and Information Theory. Edited by M. Behara, K. Krickeberg and J. Wolfowitz. IV, 256 pages. 1969. DM 18,– / $ 5.00

Vol. 90: N. P. Bhatia and O. Hajek, Local Semi-Dynamical Systems. II, 157 pages. 1969. DM 14,– / $ 3.90

Vol. 91: N. N. Janenko, Die Zwischenschrittmethode zur Lösung mehrdimensionaler Probleme der mathematischen Physik. VIII, 194 Seiten. 1969. DM 16,80 / $ 4.70

Vol. 92: Category Theory, Homology Theory and their Applications II. Edited by P. Hilton. V, 308 pages. 1969. DM 20,– / $ 5.50

Vol. 93: K. R. Parthasarathy, Multipliers on Locally Compact Groups. III, 54 pages. 1969. DM 5,60 / $ 1.60

Vol. 94: M. Machover and J. Hirschfeld, Lectures on Non-Standard Analysis. VI, 79 pages. 1969. DM 6,– / $ 1.70

Vol. 95: A. S. Troelstra, Principles of Intuitionism. II, 111 pages. 1969. DM 10,– / $ 2.80

Vol. 96: H.-B. Brinkmann und D. Puppe, Abelsche und exakte Kategorien, Korrespondenzen. V, 141 Seiten. 1969. DM 10,– / $ 2.80

Vol. 97: S. O. Chase and M. E. Sweedler, Hopf Algebras and Galois theory. II, 133 pages. 1969. DM 10,– / $ 2.80

Vol. 98: M. Heins, Hardy Classes on Riemann Surfaces. III, 106 pages. 1969. DM 10,– / $ 2.80

Vol. 99: Category Theory, Homology Theory and their Applications III. Edited by P. Hilton. IV, 489 pages. 1969. DM 24,– / $ 6.60

Vol. 100: M. Artin and B. Mazur, Etale Homotopy. II, 196 Seiten. 1969. DM 12,– / $ 3.30

Vol. 101: G. P. Szegö et G. Treccani, Semigruppi di Trasformazioni Multivoche. VI, 177 pages. 1969. DM 14,– / $ 3.90

Vol. 102: F. Stummel, Rand- und Eigenwertaufgaben in Sobolewschen Räumen. VIII, 386 Seiten. 1969. DM 20,– / $ 5.50

Vol. 103: Lectures in Modern Analysis and Applications I. Edited by C. T. Taam. VII, 162 pages. 1969. DM 12,– / $ 3.30

Vol. 104: G. H. Pimbley, Jr., Eigenfunction Branches of Nonlinear Operators and their Bifurcations. II, 128 pages. 1969. DM 10,– / $ 2.80

Vol. 105: R. Larsen, The Multiplier Problem. VII, 284 pages. 1969. DM 18,– / $ 5.00

Vol. 106: Reports of the Midwest Category Seminar III. Edited by S. Mac Lane. III, 247 pages. 1969. DM 16,– / $ 4.40

Vol. 107: A. Peyerimhoff, Lectures on Summability. III, 111 pages. 1969. DM 8,– / $ 2.20

Vol. 108: Algebraic K-Theory and its Geometric Applications. Edited by R. M. F. Moss and C. B. Thomas. IV, 86 pages. 1969. DM 6,– / $ 1.70

Vol. 109: Conference on the Numerical Solution of Differential Equations. Edited by J. Ll. Morris. VI, 275 pages. 1969. DM 18,– / $ 5.00

Vol. 110: The Many Facets of Graph Theory. Edited by G. Chartrand and S. F. Kapoor. VIII, 290 pages. 1969. DM 18,– / $ 5.00

Vol. 111: K. H. Mayer, Relationen zwischen charakteristischen Zahlen. III, 99 Seiten. 1969. DM 8,– / $ 2.20

Vol. 112: Colloquium on Methods of Optimization. Edited by N. N. Moiseev. IV, 293 pages. 1970. DM 18,– / $ 5.00

Vol. 113: R. Wille, Kongruenzklassengeometrien. III, 99 Seiten. 1970. DM 8,– / $ 2.20

Vol. 114: H. Jacquet and R. P. Langlands, Automorphic Forms on GL (2). VII, 548 pages. 1970. DM 24,– / $ 6.60

Vol. 115: K. H. Roggenkamp and V. Huber-Dyson, Lattices over Orders I. XIX, 290 pages. 1970. DM 18,– / $ 5.00

Vol. 116: Sèminaire Pierre Lelong (Analyse) Année 1969. IV, 195 pages. 1970. DM 14,– / $ 3.90

Vol. 117: Y. Meyer, Nombres de Pisot, Nombres de Salem et Analyse Harmonique. 63 pages. 1970. DM 6.– / $ 1.70

Vol. 118: Proceedings of the 15th Scandinavian Congress, Oslo 1968. Edited by K. E. Aubert and W. Ljunggren. IV, 162 pages. 1970. DM 12,– / $ 3.30

Vol. 119: M. Raynaud, Faisceaux amples sur les schémas en groupes et les espaces homogènes. III, 219 pages. 1970. DM 14,– / $ 3.90

Vol. 120: D. Siefkes, Büchi's Monadic Second Order Successor Arithmetic. XII, 130 Seiten. 1970. DM 12,– / $ 3.30

Vol. 121: H. S. Bear, Lectures on Gleason Parts. III, 47 pages. DM 6,–/$ 1.70

Vol. 122: H. Zieschang, E. Vogt und H.-D. Coldewey, Flächen ebene diskontinuierliche Gruppen. VIII, 203 Seiten. 1970. DM 1 $ 4.40

Vol. 123: A. V. Jategaonkar, Left Principal Ideal Rings. VI, 145 p 1970. DM 12,– / $ 3.30

Vol. 124: Séminare de Probabilités IV. Edited by P. A. Meyer. IV, pages. 1970. DM 20,– / $ 5.50

Vol. 125: Symposium on Automatic Demonstration. V, 310 pages. DM 20,– / $ 5.50

Vol. 126: P. Schapira, Théorie des Hyperfonctions. XI, 157 pages. DM 14,– / $ 3.90

Vol. 127: I. Stewart, Lie Algebras. IV, 97 pages. 1970. DM 10,– / $

Vol. 128: M. Takesaki, Tomita's Theory of Modular Hilbert Alg and its Applications. II, 123 pages. 1970. DM 10,– / $ 2.80

Vol. 129: K. H. Hofmann, The Duality of Compact Semigroup C*- Bigebras. XII, 142 pages. 1970. DM 14,– / $ 3.90

Vol. 130: F. Lorenz, Quadratische Formen über Körpern. II, 77S 1970. DM 8,– / $ 2.20

Vol. 131: A Borel et al., Seminar on Algebraic Groups and Re Finite Groups. VII, 321 pages. 1970. DM 22,– / $ 6.10

Vol. 132: Symposium on Optimization. III, 348 pages. 1970. DM 2 $ 6.10

Vol. 133: F. Topsøe, Topology and Measure. XIV, 79 pages. 1 DM 8,– / $ 2.20

Vol. 134: L. Smith, Lectures on the Eilenberg-Moore Spectral Sequ VII, 142 pages. 1970. DM 14,– / $ 3.90

Vol. 135: W. Stoll, Value Distribution of Holomorphic Maps into pact Complex Manifolds. II, 267 pages. 1970. DM 18,– / $

Vol. 136: M. Karoubi et al., Séminaire Heidelberg-Saarbrücken-buorg sur la K-Théorie. IV, 264 pages. 1970. DM 18,– / $ 5.00

Vol. 137: Reports of the Midwest Category Seminar IV. Edit S. MacLane. III, 139 pages. 1970. DM 12,– / $ 3.30

Vol. 138: D. Foata et M. Schützenberger, Théorie Géométriqu Polynômes Eulériens. V, 94 pages. 1970. DM 10,– / $ 2.80

Vol. 139: A. Badrikian, Séminaire sur les Fonctions Aléatoires aires et les Mesures Cylindriques. VII, 221 pages. 1970. DM 1 $ 5.00

Vol. 140: Lectures in Modern Analysis and Applications II. Edi C. T. Taam. VI, 119 pages. 1970. DM 10,– / $ 2.80

Vol. 141: G. Jameson, Ordered Linear Spaces. XV, 194 pages. DM 16,– / $ 4.40

Vol. 142: K. W. Roggenkamp, Lattices over Orders II. V, 388 p 1970. DM 22,– / $ 6.10

Vol. 143: K. W. Gruenberg, Cohomological Topics in Group Th XIV, 275 pages. 1970. DM 20,– / $ 5.50

Vol. 144: Seminar on Differential Equations and Dynamical Sy II. Edited by J. A. Yorke. VIII, 268 pages. 1970. DM 20,– / $

Vol. 145: E. J. Dubuc, Kan Extensions in Enriched Category T XVI, 173 pages. 1970. DM 16,– / $ 4.40

Vol. 146: A. B. Altman and S. Kleiman, Introduction to Grother Duality Theory. II, 192 pages. 1970. DM 18,– / $ 5.00

Vol. 147: D. E. Dobbs, Cech Cohomological Dimensions for mutative Rings. VI, 176 pages. 1970. DM 16,– / $ 4.40

Vol. 148: R. Azencott, Espaces de Poisson des Groupes Local Compacts. IX, 141 pages. 1970. DM 14,– / $ 3.90

Vol. 149: R. G. Swan and E. G. Evans, K-Theory of Finite Grou Orders. IV, 237 pages. 1970. DM 20,– / $ 5.50

Vol. 150: Heyer, Dualität lokalkompakter Gruppen. XIII, 372 1970. DM 20,– / $ 5.50

Vol. 151: M. Demazure et A. Grothendieck, Schémas en Gro (SGA 3). XV, 562 pages. 1970. DM 24,– / $ 6.60

Vol. 152: M. Demazure et A. Grothendieck, Schémas en Grou (SGA 3). IX, 654 pages. 1970. DM 24,– / $ 6.60

Vol. 153: M. Demazure et A. Grothendieck, Schémas en Grou (SGA 3). VIII, 529 pages. 1970. DM 24,– / $ 6.60

Vol. 154: A. Lascoux et M. Berger, Variétés Kähleriennes Com VII, 83 pages. 1970. DM 8,– / $ 2.20